15th International Colloquium on Paratuberculosis (ICP 2022)

Dublin, Ireland
12-16 June 2022

ISBN: 978-1-7138-5232-2

Printed from e-media with permission by:

Curran Associates, Inc.
57 Morehouse Lane
Red Hook, NY 12571

Some format issues inherent in the e-media version may also appear in this print version.

Printed by Curran Associates, Inc. (2022)

International Association for Paratuberculosis
2015 Linden Drive
Madison, Wisconsin 53706-1102
USA

Phone: (608) 262-8457
Fax: (608) 265-6463

www.paratuberculosis.org

Additional copies of this publication are available from:

Curran Associates, Inc.
57 Morehouse Lane
Red Hook, NY 12571 USA
Phone: 845-758-0400
Fax: 845-758-2633
Email: curran@proceedings.com
Web: www.proceedings.com

TABLE OF CONTENTS

PATHOGENOMICS GENOTYPING AND MAP DIVERSITY INVITED SPEAKER

Invited Speaker for Pathogenomics Genotyping and Map Diversity 1
M. Price-Carter

PATHOGENOMICS GENOTYPING AND MAP DIVERSITY AM SESSION (1)

Investigating the Initial Interaction Between MAP and the Host Using Bovine Intestinal Organoid Models 2
R. Blake, K. Jensen, O. Alfituri, J. Hope, J. Stevens

Evidence for Local and International Spread of Mycobacterium Avium Subspecies Paratuberculosis Through Whole Genome Sequencing of Isolates from the Island of Ireland 3
V. Perets, A. Allen, J. Crispell, S. Cassidy, A. O'Connor, D. Farrell, J. Browne, J. O'Mahony, R. Skuce, K. Kenny, S. Gordon

Examining the Evolutionary Trajectories of Mycobacterium Avium Subsp. Paratuberculosis in US Dairy Cattle Using Dimensionality Reduction Techniques 4
S. Wells, Y. Wang

Involvement of M. Avium Ssp. Paratuberculosis Specific Genomic Regions LSP14/15 in Zinc Homeostasis 5
E. Goethe, K. Laarmann, J. Luhrs, A. Gieseke, R. Goethe

Genetic Features of Mycobacterium Avium Subsp. Paratuberculosis Strains Circulating in the West of France Deciphered by Whole-Genome Sequencing 6
C. Conde, T. Cochard, M. Branger, M.-N. Rossignol, C. Fourichon, A. Delafosse, A. Davergne, A. Joly, D. Ngwa-Mbot, L. Journaux, R. Guatteo, L. Schibler, J. Bannantine, F. Biet

PATHOGENOMICS GENOTYPING AND MAP DIVERSITY AM SESSION (2)

Phylogenetic Analysis of Mycobacterium Avium Subsp. Paratuberculosis in Australia 7
R. Hodgeman, R. Mann, K. Savin, N. Djitro, S. Rochfort, B. Rodoni

Understanding the Role of Mycobacterium Avium Subsp. Paratuberculosis Pathogenicity Islands by Generating Gene Knockouts Using ORBIT 8
S. Duffy, M. Behr

85 Complex Proteins as an Important Adhesion Factor of Mycobacterium Avium Subsp. Paratuberculosis in Bovine Sperm 9
J. Caldeira, M. Moreira, E. Diaz-Miranda, J. Guimaraes, Y.-F. Chang, T. Zilch, W. Barbosa, A. Junior, J. Pena

Changes in the Fecal Microbiota of Calves Experimentally Infected with Mycobacterium Avium Subsp. Paratuberculosis. 10
M.-E. Fecteau, D. Beiting, R. Sweeney

PATHOGENOMICS, GENOTYPING AND MAP DIVERSITY POSTER PITCHES

Bio-Load and Bio-type Profile of Mycobacterium Avium Subspecies Paratuberculosis Isolates Recovered from Sheep and Goats Located in Himalayan Region 11
A. Bhat, H. Malik, A. Mir, T. Hussain, K. Chaubey, S. Gupta, S. Singh

Whole-Genome Analysis of Mycobacterium Avium Subsp. Paratuberculosis IS900 Distribution Reveals Strain Type-specific Modalities 12
C. Conde, M. Price-Carter, T. Cochard, M. Branger, K. Stevenson, R. Whittington, J. Bannantine, F. Biet

Deletion in Mce4 Gene Exacerbates Virulence of Mycobacterium Avium Subsp. Paratuberculosis........ 13
M. Olivieri, R. Moyano, A. Nagel, M. Romano, N. Alonso

A Streptomycin Pre-Treatment Mouse Model of Mycobacterium Avium Subsp. Paratuberculosis Infection 14
S. Duffy, A. Lupien, F. McIntosh, M. Behr

CONTROL PROGRAMS AND EDUCATION INVITED SPEAKER

Invited Speaker for Control Programs and Education 15
H. Barkema

CONTROL PROGRAMS AND EDUCATION PM SESSION (1)

Part 3: UK Approach to Johne's Disease Control -Top Tips for Effective Control 16
P. Orpin, D. Sibley

Dynamics of MAP Infections in Small Structured Cattle Farms 17
J. Khol, T. Liening-Ewert, L. Aubock, P. Ortner, J. Kossler, A. Tichy, M. Dunser, C. Mader

Predicting Positive ELISA Results in Dairy Herds with a Preferred Herd Status in a Milk Quality Assurance Programme for Paratuberculosis 18
M. Weber, M. Aalberts, T. Dijkstra, Y. Schukken

The Irish Johne's Control Programme: A Qualitative Approach to Understanding the Motivations for Farmer and Veterinary Practitioner Engagement and Providing Information Effectively Through Co-Design to Improve Recruitment, Adherence, and Retention 19
A. Burrell, M. Good, C. McAloon, D. Graham, S. More, L. Gavey

Part 1: UK Approach to JD Control -Farmer Engagement Using Low Cost Surveillance, Risk Assessment and Prevalence Prediction 20
P. Orpin, D. Sibley

CONTROL PROGRAMS AND EDUCATION PM SESSION (2)

Changes in Johne's Disease Control Practices in Ontario Canada Between 2013 and 2019 21
J. Imada, S. Roche, C. Bauman, D. Kelton

Raising Awareness on Paratuberculosis in Africa 22
J. Okuni, J. Ssekitoleko, K. Eltom, W. Elmagzoub, S. Idris, E. Eltayeb, A. Amanzada, J. Omega, F. Tanneberger, L. Ojok, A. Wahed

Different Mycobacterium Avium Subspecies Paratuberculosis Burden Implies Different Aims of Johnes' Control Strategies in the Three Main Dairy Production Regions of Germany 23
K. Donat, M. Krieger, A. Campe, I. Lorenz, S. Eisenberg

CONTROL PROGRAMMES AND EDUCATION POSTER PITCHES

Completing Herd Investigations Using a Mobile App with Links to Real-Time Data 24
L. Citer, L. Gavey, P. O'Sullivan, D. Graham

A Preliminary Analysis of vRAMP Responses for Dairy Herds Entering the Voluntary Johne's Disease Control Programme in Northern Ireland During Its First Year of Operation 25
K. Meek, S. Strain, N. O'Connell, I. Grant

Irish Johne's Control Programme Description and Progress 26
L. Gavey, S. More, L. Citer, D. Graham

Important Role for Abattoir Monitoring and Vaccination in Control Programs for Bovine Paratuberculosis Due to Mycobacterium Avium Subsp. Paratuberculosis (Map) 27
I. Links, H. Dahmen

How Multiplex Diagnostics Can Help in the Rapid Screening of Johne's Disease: A Proposed Surveillance Method 28
N. Gordon, J. Tejerina, D. Whittaker, H. Moore, R. Janeczko

Investigating the Dynamics of Johne's Disease in Dairy Herds After a Province Wide Voluntary Control Program in Ontario Canada 29
J. Imada, S. Roche, C. Bauman, D. Kelton

DIAGNOSTICS AND DETECTION INVITED SPEAKER

Invited Speaker for Diagnostics and Detection 30
V. Kapur

DIAGNOSTICS AND DETECTION AM SESSION (1)

Comparison of Herd-Level MAP Prevalence Based on F57 and ISMAP02 Target Genes 31
L. Martins, K. Orsel, J. De Buck, H. Barkema

Evaluation of a Droplet Digital PCR (ddPCR) Assay for Detection and Quantification of Mycobacterium Avium Subsp. Paratuberculosis (MAP) DNA in Whole-Blood and Fecal Samples from MAP-infected Holstein Cattle 32
G. Badia-Bringue, M. Canive, R. Casais, C. Blanco-Vazquez, J. Amado, N. Iglesias, R. Juste, M. Alonso

Evaluation of the Actiphage® Rapid JD Test as a Pre-Breeding Heifer Screening Tool in Endemically Infected Herds 33
T. Geraghty, G. Caldow, G. Richardson

Development of LAMP Coupled LFD Kit for Rapid and Specific Detection of Mycobacterium Avium Subspecies Paratuberculosis in Field 34
J. Sohal, M. Singh, M. Jain, S. Singh, A. Singh, S. Gupta, J. Reddy, R. Polavarapu, A. Kumar, D. Sharma, G. Aseri

Evaluation of the Bovine ATP-Binding Cassette Subfamily a Member 13 (ABCA13) as a Potential Biomarker for Sensitive Detection of Animals with Focal Pathological Forms of Subclinical Paratuberculosis.......... 35
R. Casais, C. Blanco-Vazquez, N. Iglesias, P. Vazquez, R. Juste, J. Garrido, A. Balseiro, M. Canive, J. Amado, M. Queipo, T. Iglesias, M. Alonso-Hearn

DIAGNOSTICS AND DETECTION AM SESSION (2)

A Novel 2-In-1 Culture System for the Isolation of Mycobacterium Avium Subsp. Paratuberculosis 36
H. Dane, L. Stewart, I. Grant

More Insights About the Efficacy of Copper Ion Treatment on Mycobacterium Avium Subsp. Paratuberculosis (MAP). a Clue for the Observed Tolerance.......... 37
C. Tejeda, M. Villegas, P. Steuer, . Salgado

Serum Metabolome Changes in Bovine Paratuberculosis by NMR Spectroscopy and Blood Indices.......... 38
N. Pozzato, A. Massaro, A. Tata, I. Pallante, M. Bottazzari, P. Fiorini, M. Pra, A. Stefani, V. Bertazzo, J. De Buck, R. Piro

The Fate of Heifers with 'low-Positive' Milk ELISA Results 39
M. Weber, M. Aalberts, T. Dijkstra, Y. Schukken

Development of a Rapid Multiplex Immunoassay for the Early Detection of Mycobacterium Avium Paratuberculosis in Bovine Samples.......... 40
N. Gordon, J. Tejerina, D. Whittaker, R. Janeczko

Early Detection of Viable MAP Shedding by Dairy Calves Possible Using a Rapid and Sensitive Phagomagnetic Separation (PhMS)-QPCR Method.......... 41
K. Meek, S. Strain, N. O'Connell, I. Grant

Phage-Based Detection of Viable Mycobacterium Avium Subsp. Paratuberculosis in Milk and Faeces Superior to Faecal qPCR and Serum- ELISA for Identifying Shedding Animals in a Northern Ireland Dairy Herd. 42
I. Grant, A. Foddai, E. Earle, J. Rankin, G. Watson, B. McInerney, S. Strain

DIAGNOSTICS AND DETECTION POSTER PITCHES (1)

Mycobacteriophage Characterisation: An Important Consideration for the Biocontrol of Mycobacterium Avium Sbsp. Paratuberculosis.......... 43
L. O'Connell, J. O'Mahony

Study on Prevalence of Mycobacterium Avium Subspecies Paratuberculosis (MAP) Infection in Suspected Diarrhoeic Buffaloes and Cattle Reporting at Clinics for Therapeutic Intervention 44
S. Sharma, A. Gautam, S. Singh, K. Chaubey, S. Harihahan

Detection of MAP and M. Avium Co-Infection in North American Bison Using Actiphage.......... 45
W. Hartill, C. Evans, A. Hunyh, T. Perehinec, B. Swift, C. Rees

Optimization of Actiphage for Detection of MAP in Different Hosts, Including Humans 46
B. Swift, W. Handley-Hartill, A. Kubala, C. Evans, C. Rees

HOST RESPONSE AND IMMUNOLOGY INVITED SPEAKER

Invited Speaker for Host Response and Immunology 47
Kumudika De Silva

HOST RESPONSE AND IMMUNOLOGY AM SESSION (1)

An in Vivo Vaccine Screening Model in Neonatal Calves to Investigate Parenteral Vaccine-Induced Mucosal Immune Responses to an Enteric MAP Infection 48
A. Facciuolo, A. Lee, N. Rawlyk, S. Napper, A. Potter, V. Gerdts, H. Townsend, R. Hancock, P. Griebel

A Genome-Wide Association Study for Tolerance to Bovine Paratuberculosis Identifies Genetic Regions with Candidate Genes Involved in DNA Packaging, DNA Damage Response and Innate Immunity 49
M. Alonso, M. Canive, G. Badia-Bringue, P. Vazquez, O. Gonzalez-Recio, A. Fernandez, J. Garrido, R. Juste

Dynamics of Immune Responses Following Immunization with a Live Attenuated Vaccine Against Johne's Disease. 50
A. Talaat, M. Hanafy

Regionally Distinct Mucosal Immune Responses to MAP Infection in Discrete and Continuous Peyer's Patches of Young Calves at 2 and 12 Months Post- Infection 51
A. Facciuolo, A. Lee, P. Cano, S. Napper, A. Potter, V. Gerdts, R. Hancock, L. Mutharia, P. Griebel

Characterising Immune Responses Following Vaccination of Calves with Live- Attenuated Strains of Mycobacterium Avium Subsp. Paratuberculosis 52
R. Samani, J. De Buck

HOST RESPONSE AND IMMUNOLOGY AM SESSION (2)

Identification of Genetic Regions Associated with a Low Mycobacterium Avium Subsp. Paratuberculosis Load Within Infected Monocyte-Derived Macrophages 53
M. Canive, G. Badia-Bringue, M. Alonso-Hearn

Understanding the Invasion Dynamics of Mycobacterium Avium Subsp. Paratuberculosis (MAP) 54
G. Baruta, H. Zhang, L. Alston, J. De Buck, S. Hirota

Uncovering miRNA Signatures of Resilience to Paratuberculosis Infection and Their Functional Relevance in Disease Progression 55
K. Wright, K. De Silva, A. Purdle, K. Plain, S. Oehlers

Alternative Splicing of pre-MRNA Modulates the Immune Response in Peripheral Blood and Gut Tissues of Holstein Cattle Naturally Infected with Mycobacterium Avium Subsp. Paratuberculosis 56
G. Badia-Bringue, M. Canive, J. Lavin, R. Casais, C. Blanco-Vazquez, M. Alonso-Hearn

Identification of Disease Tolerant and Susceptible Cows to Mycobacterium Avium Subspecies Paratuberculosis Infection for Genetic Analyses of Disease Resistance 57
N. Bissonnette, D. Kelton, G. Fecteau, K. Tahlan, E. Ibeagha-Awemu, P. Griebel

HOST RESPONSE AND IMMUNOLOGY POSTER PITCHES

An Integrative Mendelian Randomization Analysis of Genome-Wide Association and eQTLs Studies Identified Novel Genes Associated with Bovine Paratuberculosis Susceptibility 58
G. Badia-Bringue, M. Canive, N. Fernandez, P. Vazquez, J. Lavin, J. Bilbao, R. Casais, C. Blanco-Vazquez, J. Garrido, R. Juste, M. Alonso-Hearn

Peripheral Cellular Immune Responses in Mycobacterium Avium Subspecies Paratuberculosis (MAP) Infected and Vaccinated Goats: A Comparative Study 59
S. Sharma, A. Gautam, S. Singh, K. Chaubey, S. Gupta

DIAGNOSTICS AND DETECTION POSTER PITCHES (2)

Validation of a IS900-QPCR Assay for the Detection of Paratuberculosis in Faeces According to the OIE "Principles and Methods of Validation of Diagnostic Assays for Infectious Disease". 60
S. Russo, G. Galleti, A. Filippi, N. Arrigoni, C. Garbarino, M. Ricchi

Bayesian Accuracy Estimates of Environmental Sampling for Determining Herd Paratuberculosis Infection Status and Its Association with the Within- Herd Prevalence in Québec Dairies 61
J. Sabogal, G. Fecteau, E. Dore, G. Cote, J.-P. Roy, V. Wellemans, S. Buczinski

Identification of Mycobacterium Avium Subsp. Paratuberculosis Using a Single-Chain Antibody Targeting the Cell Wall Lipopentapeptide L5P 62
H. Bach, Y. Slavin, A. Gonzalez, C. Ganneau, S. Bay, F. Biet

Evaluating the Performance of Various Protocols for the Diagnostic Performance in Paratuberculosis IS900 PCR 63
M. Jain, H. Gandhi, S. Mourya, A. Kumar, G. Aseri, D. Sharma, J. Singh

Herd Environmental Sampling for Detection of Mycobacterium Avium Subspecies Paratuberculosis in Pasture-Based Dairy Herds – Preliminary Results 64
N. Field, J. Mee, C. McAloon

EPIDEMIOLOGY AND ECONOMICS INVITED SPEAKER

Invited Speaker Epidemiology and Economics 65
F. Griffin

EPIDEMIOLOGY AND ECONOMICS AM SESSION (1)

Dam and Environmental Transmission Routes on Mycobacterium Avium Paratuberculosis Infection in Calves 66
E. Nunney, M. Crotta, K. Bond, M. Green, S. Van Winden, J. Guitian

Economic Premiums Associated with Mycobacterium Avium Subspecies Paratuberculosis-Negative Replacement Purchases in Major Dairy Producing Regions. 67
P. Rasmussen, H. Barkema, E. Beaulieu, S. Mason, D. Hall

Effects of Silirum® Based Vaccination Programs on Map Faecal Shedding and Serological Response in Seven French Dairy Herds. 68
F. Corbiere, D. Guellouz, J. Cournet, C. Tasca, E. Dubaux, L. Foures, G. Foucras

Insights on Paratuberculosis in Cattle at a Regional Scale Using Data-Driven Modeling and Inference 69
G. Beaunee, P. Ezanno, A. Joly, P. Nicolas, E. Vergu

Success of Paratuberculosis Control is Associated with Calf Mortality Rate in Dairy Herds 70
A. Klassen, S. Sollner-Donat, E. Einax, K. Donat

EPIDEMIOLOGY AND ECONOMICS AM SESSION (2)

Mycobacterium Avium Subspecies Paratuberculosis (MAP) Elisa Tests on Milk to Predict Future Health and Culling in a Dairy Herd 71
R. Sibley, P. Orpin

Exposure to Mycobacterium Avium Subsp. Paratuberculosis in Alpine Pastures (Northern Italy): Evaluation of Cattle and Red Deer (Cervus Elaphus) Contribution Through Environmental Faecal Samples. 72
A. Filippi, C. Luzzago, M. Nava, A. Forti, S. Russo, F. Barsi, L. Corlatti, L. Pendrotti, A. Bianchi, M. Ricchi, . Ricchi, N. Arrigoni, C. Garbarino

Factors Associated with the Spread of Paratuberculosis in Hungarian Dairy Cattle Herds 73
V. Jurkovich, M. Bakony, B. Vass-Bognar

Multi-Institutional Investigation for Determining the Status of Johne's Disease in the Suspected and Apparently Healthy Population of Domestic Livestock Species: A Pan-India Study 74
S. Singh, D. Chauhan, J. Sohal, K. Gunaseelan, A. Singh, K. Porteen, K. Chaubey, S. Gupta, P. Singh

Epidemiology of Johne's Disease in the New Zealand Farmed Deer Industry 1980-2020 75
S. Norton, M. Price-Carter, F. Griffin

EPIDEMIOLOGY AND ECONOMICS POSTER PITCHES

Seroprevalence of Mycobacterium Avium Subsp. Paratuberculosis in Austrian Goat Flocks 76
J. Schrott, E. Sodoma, M. Dunser, A. Tichy, J. Khol

Understanding the Antibacterial Mechanisms of Copper Ion Treatment on Mycobacterium Avium Subsp. Paratuberculosis 77
C. Tejeda, M. Villegas, P. Steuer, M. Salgado

Molecular and Serological Survey of Mycobacterium Avium Subspecies Paratuberculosis in Ugandan Cattle 78
J. Ssekitoleko, L. Ojok, E. Eltyayeb, A. Amanzada, K. Eltom, A. Wahed, J. Okuni

Johne's Disease in New Zealand; the Study of a Milk Antibody Dataset 79
R. White, S. Pryor, R. Thresher

PUBLIC HEALTH AND MAP IN THE ENVIRONMENT INVITED SPEAKER

Invited Speaker for Public Health and Map in the Environment 80
M. Behr

PUBLIC HEALTH AND MAP IN THE ENVIRONMENT PM SESSION (1)

Heat Resistance of Mycobacterium Avium Subsp. Paratuberculosis: Inactivation Kinetics During the Production Process of Mozzarella Cheese 81
F. Barsi, E. Cosciani-Cunico, S. Russo, E. Dalzini, A. Filippi, G. Cammi, N. Arrigoni, P. Daminelli, N. Losio, C. Garbarino, M. Ricchi

Effluent as a Herd-Level Screening Tool for the Estimation of Johne's Disease Prevalence Within New Zealand Milking Herds 82
R. White, S. Pryor, R. Thresher

Presence of Mycobacterium Avium Spp. Paratuberculosis in Silage Collected on Commercial Cattle Farms 83
S. Eisenberg, A. Campe, K. Donat

PUBLIC HEALTH AND MAP IN THE ENVIRONMENT PM SESSION (2)

Detection of Mycobacterium Avium Subspecies Paratuberculosis in Environmental Samples from a Dairy Goat Farm in Germany 84
C. Pickrodt, K. Donat, H. Kohler

Environmental Free-Living Amoebae Interaction with Mycobacterium Avium Subsp. Paratuberculosis 85
A. Jessu, T. Cochard, A. Samba-Louaka, J.-L. Moyen, F. Biet, Y. Hechard

Cumulative Sensitivity and Specificity of Repeated Environmental Sampling and Milk-Pool Testing as Surveillance Methods at Herd Level in Large Low- Prevalence Dairy Herds 86
K. Donat, A. Wichert, E. Einax, E. Kasbohm

Therapeutic Management of Mycobacterium Avium Subspecies Paratuberculosis Infection with Complete Resolution of Disease in Patients with Inflammatory Bowel Syndrome in India 87
S. Gupta, K. Chaubey, P. Agarwal, A. Gautam, M. Singh, P. Singh, S. Singh

ADDITIONAL PAPERS

Combining Optimized Sample Extraction with Sensitive Detection Using Bactotype® MAP Real-Time PCR 88
N. Tandon, C. Gaunitz, M. Labitzke, C. Engemann, O. Sasse, L. Moussi, F. Ullman

Evaluation of Fecal Culture, Fecal RT-PCR, Three Commercially Available Serum ELISA's, Complement Fixation and AGID Assays as Diagnostic Tools to Detect Mycobacterium Avium Ssp. Paratuberculosis in Dairy Sheep 89
J. Cooper, J. Smith, N. Aulik

Detection of a Mycobacterium Avium Spp Avium in a Pudu and Muntjac in the United States 90
J. Cooper, A. Talaat, A. Mutlow, K. Conley, N. Aulik

Detection of Antibodies Against Mycobacterium Avium Subspecies Paratuberculosis in Sera and Bulk Tank Milk Samples from French Goat Herds. 91
V. Thibault-Poisson, M. Thirion, L. Pineau, I. Bremaud, M. Tabouret

Interactions Between Mycobacterium Avium Subsp. Paratuberculosis and Staphylococcus Aureus and Streptococcus Agalactiae, in Epithrlial Cells of the Bovine Mammary Gland - Mac-T 92
J. Pena, D. Schwarz, A. Brasil, L. Oliveira, A. Silva, M. Moreira

Results of Fifteen Years Milk Quality Assurance for Paratuberculosis in Dutch Dairy Herds 93
M. Weber, M. Aalberts, T. Dijkstra, Y. Schukken

Performance of Three Commercial ELISA Tests for Mycobacterium Avium Subsp. Paratuberculosis (MAP) Antibodies in Blood and Milk Samples 94
I. Lorenz

Is Copper Ions Treatment Enough to Stop Mycobacterium Avium Subsp. Paratuberculosis (MAP) Infection Progression in Newborn Calves? 95
M. Salgado, P. Steuer, C. Tejea, M. Moroni, C. Verdugo

Twenty-Five Topics to Be Considered in the Design of Paratuberculosis Control Programmes for Dairy Herds 96
M. Weber, T. Dijkstra, M. Aalberts, Y. Schukken

The Biosecurity and Bio-Containment Risks Associated with MAP Infection and Prevalence in UK Dairy Herds 97
R. Sibley, P. Orpin

Prediction of Clinical Paratuberculosis with RT-QPCR 98
B. Vass-Bognar, W. Baumgartner, K. Fornyos, V. Jurkovich

Evaluation of Single Nucleotide Polymorphisms (SNPs) Associated to Genetic Resistance to Paratuberculosis in Marchigiana Beef Cattle 99
P. Mazzone, L. Petrucci, M. Torricelli, S. Corneli, A. Di Paolo, C. Sebastiani, M. Ciullo, M. Sebastianelli, S. Costarelli, E. Scoccia, F. Sbarra, F. Gabbianelli, A. Valentini, G. Filippini, G. Pezzotti, M. Biagetti

Humoral Immune Response Against Mycobacterium Avium Subspecies Paratuberculosis Immunogenic Epitopes in Type 1 Diabetes Pediatric Patients 100
M. Noli, G. Meloni, D. Cossu, S. Hossein, S. Jasemi, L. Sechi

Clinical Trial (phase II) of Vaccination with a Heat-Killed Local Strain of Mycobacterium Avium Subsp. Paratuberculosis 101
R. Moyano, M. Olivieri, A. Nagel, X. Cuerda, M. Gravisaco, F. Pinedo, P. Sosa, G. Traveria, M. Alonso, M. Romano

Evaluation of a Direct Multiplex qPCR Assay and ELISA for the Diagnosis of Bovine Paratuberculosis in a Tuberculosis Endemic Area 102
F. Jurado-Martos, I. Luque, C. Tarradas, B. Huerta, F. Cardoso

Association Between Bovine MHC II Haplotypes and Susceptibility Traits to Bovine Johne's Disease 103
V. Garcia, R. Whittington, A. Thomas, C. Davies, J.-A. Toribio, K. Plain

Probiotic Bacteria Modulate Immune Responses to Paratuberculosis Vaccination 104
M. Oyanguren, E. Molina, M. Mujika, A. Badiola, I. Ladero, N. Elguezabal

Optimization and Validation of Lateral Flow Assay for Rapid Diagnosis of Paratuberculosis 105
M. Jain, M. Singh, A. Singh, S. Gupta, R. Polavarapu, A. Kumar, G. Aseri, D. Sharma, J. Sohal

Recombinant Johnin (rJohnin) Assay for the DTH Based Specific Diagnosis of Paratuberculosis 106
M. Jain, A. Kumar, M. Singh, S. Gupta, R. Polavarapu, V. Singh, A. Kumar, D. Sharma, G. Aseri, J. Sohal

Penside Plate rELISA as 'Herd Screening Assay' for Onsite Diagnosis of Paratuberculosis 107
M. Jain, A. Singh, S. Gupta, R. Polavarapu, A. Kumar, D. Sharma, G. Aseri, J. Sohal

A Team-Based Approach to Farmer Engagement in Johne's Disease Control 108
L. Citer, L. Gavey, D. Graham, E. Farrell, S. McElroy

Direct qPCR is Equally Sensitive as Liquid Culture for the Detection of MAP in Ovine Faecal Samples 109
H. Koehler, C. Pickrodt, P. Mobius

Exploring the Best Peptide Regions of Secretome of Mycobacterium Avium Subspecies Paratuberculosis for Better Functionality and Binding Efficacy to Design Peptides Based ELISA................ 110
S. Hariharan, K. Chaubey, P. Singh, S. Singh

No Evidence of Transmission of Mycobacterium Avium Subspecies Paratuberculosis from the Udder Skin into the Colostrum of Dairy Goats 111
C. Pickrodt, K. Donat, U. Moog, H. Kohler

Evaluation of Four Tests (screening and Confirmatory) for the Presence of MAP Infection in the Spiked and Un-Spiked Milk Samples.. 112
K. Chaubey, M. Singh, S. Gupta, P. Singh, A. Singh, D. Chauhan, S. Singh

Understanding the Antibacterial Mechanisms of Copper Ion Treatment on Mycobacterium Avium Subsp. Paratuberculosis.. 113
C. Tejeda, M. Villegas, P. Steuer, M. Salgado

A Novel Paraffin Slide Culture (PSC) Technique to Isolate and Characterize Non-Tuberculosis Mycobacteria (NTM).. 114
M. Jain, S. Maurya, R. Narang, P. Narang, S. Rana, A. Kumar, G. Aseri, D. Sharma, J. Singh

Search for Bacterial Gut Microbiome Patterns and Factors Associated with Pathogenesis of Paratuberculosis in Goats 115
P. Moebius, E. Liebler-Tenorio, H. Koehler, C. Pickrodt, T. Fuchs

Development of a Diagnostic Test to Identify MAP in the Blood and Tissue of Patients with Crohn's Disease.. 116
N. Rayment, A. Hermon-Taylor, B. Hudspith, R. Pickup, G. Rhodes, G. Agrawal, J. Sanderson, J. Hermon-Taylor

Developing a Novel Johne's Progress Tracker to Support the UK National Management Plan for Dairy Herds 117
P. Orpin, N. Taylor, D. Sibley

Part 2: UK Approach to Johne's Disease Control -Utilising Web-based Tools to Enhance Farmer Engagement.. 118
P. Orpin, D. Sibley

Achieving 93% Uptake with UK National Johne's Management Plan Using Commercial Drivers and Industry Support 119
P. Orpin, D. Sibley

Performance Evaluation of an Indirect milk-ELISA for the Diagnosis of Mycobacterium Avium Subsp. Paratuberculosis in Individual Samples of Dairy Sheep 120
A.-R. Attili, E. Bonacucina, M. Linardi, L. Pacifici, V. Ngwa, V. Cuteri

In Contrast to Successfully Detecting Viable MAP from Blood Using a phage/qPCR Assay Within 48 Hours, by Two Years Routine Mycobacterial Blood Culture Provides No Clinically Useful Information 121
R. Greenstein, L. Su, J. Stabel, I. Grant, A. Foddai, S. Brown

Proteomic Analysis of Sera from Holstein Friesian Cows with Different Pathological Forms of Bovine Paratuberculosis (PTB) 122
A. Leon, M. Munoz, N. Iglesias, C. Vazquez, A. Balseiro, F. Corrales, S. Ciordia, F. Milhano, R. Casais

Evaluation of Diagnostic Tests for Mycobacterium Avium Subsp. Paratuberculosis Following a Single Screening of Animals in Herds Endemic with Johne's Disease 123
E. Ramovic, R. Cannon, D. Yearsley, M. Griffin, D. Prendergast, T. Clegg, D. Hand, E. Nighallchoir, B. Markey, J. Egan

Longitudinal Study on Mycobacterium Avium Subsp. Paratuberculosis Antibody Kinetics and Excretion in Dairy Sheep Using Blood, Milk and Faeces Throughout the Lactation Period 124
A.-R. Attili, E. Bonacucina, M. Linardi, L. Pacifici, L. Galosi, V. Ngwa, V. Cuteri

Joint Control of Bovine Tuberculosis and Paratuberculosis – the Case for and Against........ 125
R. Whittington, S. Eisenberg, F. Griffin, A. Michel, E. Schwan, K. Stevenson, M. Webber

Isolation of Novel Lytic Bacteriophages with the Potential to Be Used for Detection of Viable Mycobacterium Avium Subsp. Paratuberculosis........ 126
H. Dane, L. Stewart, I. Grant

The Silent Sleep of Intramacrophage Mycobacterium Avium Subspecies Paratuberculosis 127
J. Stevens, K. Jensen

Factors Affecting the Likelihood of a Positive or Negative Mycobacterium Avium Subspecies Paratuberculosis (MAP) Faecal PCR Result in ELISA Seropositive Or Inconclusive Animals in Irish Herds – Preliminary Results........ 128
N. Field, J. Mee, C. McAloon

Performance of the Molecular Diagnosis Using qPCR IS900 for the Detection of Mycobaceterium Avium Subspecies Paratuberculosis in Cattle Feces in Chiriquí, Panama........ 129
F. Acosta, D. De La Hoz, C. De La Guardia, A. De Souza, D. Sambrano, A. Garcia, D. Candanedo, D. Palacios, S. Madrid, V. Polanco, A. Prichard, A. Villarreal, M. Jaen, J. Pitti, C. Nader, A. Goodridge

Single Nucleotide Polymorphisms Genotyping of Cattle from Uganda and Sudan and Their Connection to Mycobacterium Subspecies Paratuberculosis Infection in Ugandan and Sudanese Cattle 130
J. Okuni, J. Ssekitoleko, W. Elmagzoub, S. Idris, M. Mpumwiire, M. Kisekka, L. Ojok, A. Amanzada, E. Eltyayeb, K. Eltom, A. Wahed

Lessons from the Irish Johne's Control Programme 131
L. Gavey, D. Graham, S. More, L. Citer

Genetic Typing of Mycobacterium Avium Subsp. Paratuberculosis Strains Isolated from 14 Cows Shedding High Levels of MAP in the Provinces of Quebec and Ontario Herds 132
A. Byrne, K. Tahlan, S. Ollier, T. Cochard, F. Biet, N. Bissonnette

Bayesian Estimation of Sensitivity and Specificity of Fecal Culture, Fecal PCR and Serum ELISA for Diagnosis of Mycobacterium Avium Subsp. Paratuberculosis Infections in Egyptian Sheep 133
J. Sabogal, I. Elsohaby, A. Selim

Evaluation of Associations Between MAP-C Or MAP-S Genotypes and the Severity of Histo-/pathological Abnormalities of Infected Cattle, Red Deer and Sheep from Germany and New Zealand........ 134
M. Price-Carter, M. Gautam, P. Biggs, M. Weber, H. Koehler, C. Heuer, P. Mobius

Author Index

319

Invited Speaker for Pathogenomics Genotyping and Map Diversity

Dr. Marian Price-carter

Pathogenomics Genotyping and Map Diversity - Invited Speaker, Main Auditorium, Printworks, June 13, 2022, 8:30 AM - 9:15 AM

Five learning points

1. Simple test and cull strategies that have led to eradication of bovine tuberculosis from livestock in other parts of the world are hampered in New Zealand (NZ) by reinfection from a wildlife reservoir.

2. OSPRI TBfree, a farmer driven organization, has greatly reduced M. bovis herd prevalence by taking an active, all-inclusive approach where they monitor and control infection both in livestock herds and wildlife, and regulate movement of stock.

3. Different types of M. bovis are regionally localized in NZ and for multiple decades strain typing has enhanced the efficiency of control by guiding OSPRI in the movement control of livestock, control within herds and enhanced wildlife control in the area of new herd infections.

4. The single nucleotide polymorphism (SNP) lineages that result from whole genome sequencing (WGS) analyses are proving to be far superior to previously employed typing assays for guiding control efforts.

5. Information from WGS about shared common ancestors has been helpful for clearly defining relationships and movement of NZ M. bovis, M. pinnipedii and MAP types.

194

Investigating the initial interaction between MAP and the host using bovine intestinal organoid models

Miss Rosemary Blake[1], Dr. Kirsty Jensen, Dr. Omar Alfituri, Professor Jayne Hope, Dr. Jo Stevens
[1]Roslin Institute, The University Edinburgh, Edinburgh, United Kingdom

Pathogenomics Genotyping and Map Diversity - AM Session (1), Main Auditorium, Printworks, June 13, 2022, 9:15 AM - 10:30 AM

The current control measures for Johne's disease (JD), caused by Mycobacterium avium ssp paratuberculosis (MAP), are ineffective and do not reduce spread of the disease or infection. Better diagnostic tools and vaccine targets must be identified if we hope to mitigate the impact of JD on the economy and improve animal welfare. By investigating the early interaction of MAP with the host at a cellular and molecular level, new diagnostic markers and vaccine targets will be identified. Due to the slow-growing nature of MAP and its long subclinical period, studying the infection in vivo is difficult. Therefore, we have developed and utilised a bovine intestinal organoid (enteroid) system which is reproducible and physiologically relevant. Both 2D monolayers and 3D apical out enteroids were characterised using immunofluorescence microscopy and RT-qPCR to demonstrate that cell lineages were present representative of the intestine in vivo. Upon infection of these models with MAP, significantly lower numbers of the lab-adapted reference strain, K10 were recovered in comparison to a recently sampled and sequenced field isolate in our laboratory, C49. This trend was consistent between enteroid models, and may indicate MAP K10 has reduced virulence compared to more clinically relevant strains. The strain differences identified highlight the limitations of using K10 as a representative strain, which may hamper the identification of therapeutic targets in future research. The methods developed here allow an in-depth investigation of MAP infection using a physiologically representative in vitro model of the bovine intestine. This could advance our understanding of the initial host-MAP interaction with the aim of investigating more targeted diagnostic and treatment methods.

212

Evidence for local and international spread of Mycobacterium avium subspecies paratuberculosis through whole genome sequencing of isolates from the island of Ireland

Mr Viktor Perets[1], Dr Adrian Allen[2], Dr Joseph Crispell[1], Ms Sophie Cassidy[1], Dr Aoife O'Connor[3], Dr Damien Farrell[1], Dr John Browne[5], Prof Jim O'Mahony[4], Dr Robin Skuce[2], Dr Kevin Kenny[3], Prof Stephen Gordon[1,6,7,8]
[1]UCD School of Veterinary Medicine, University College Dublin, Belfield,, Ireland, [2]Agri-Food and Biosciences Institute, AFBI Stormont, Belfast,, Northern Ireland, UK, [3]Central Veterinary Research Laboratory, Department of Agriculture, Food and the Marine, Backweston,, Ireland, [4]Munster Technological University, Department of Biological Sciences, Rossa Avenue, Bishopstown,, Ireland, [5]UCD School of Agriculture and Food Science, University College Dublin, Belfield,, Ireland, [6]UCD School of Medicine, University College Dublin, Belfield,, Ireland, [7]UCD School of Biomolecular and Biomedical Science, University College Dublin, Belfield,, Ireland, [8]UCD Conway Institute, University College Dublin, Belfield,, Ireland

Pathogenomics Genotyping and Map Diversity - AM Session (1), Main Auditorium, Printworks, June 13, 2022, 9:15 AM - 10:30 AM

Whole Genome Sequencing (WGS) is increasingly being used to understand pathogen genetic diversity and tracking pathogen transmission chains. Here we describe the application of WGS to Mycobacterium avium subsp paratuberculosis, the agent of Johne's Disease in cattle, across the island of Ireland. We compare WGS to MAP diversity quantified using the standard approach of mycobacterial interspersed random unit – variable number tandem repeats (MIRU-VNTR). WGS and MIRU-VNTR were performed on a collection of 197 MAP isolates gathered from 122 cattle herds across 27 counties of the island of Ireland. Using WGS, we found the 197 isolates could be split into eight major groups while MIRU-VNTR showed lower resolution, with only two major MIRU-VNTR types.

Evidence for MAP transmission at multiple scales could be discerned from the WGS data. Dispersal across Ireland via cattle movement underscores the utility of WGS approaches to inform national control strategies. Furthermore, comparing MAP WGS data from Ireland to MAP WGS data from Great Britain and continental Europe revealed many instances of close genetic similarity and hence evidence for international transmission of infection. BEAST structured coalescent analyses estimated the rate of substitution to be 0.13 SNPs/site/year and disclosed greater transitions per lineage per year from Europe to Ireland, again indicating transmission into Ireland. Our work therefore reveals new insight into seeding of MAP infection across Ireland, highlighting how WGS can inform policy formulation to ultimately control MAP transmission at local, national and international scales.

239

Examining the evolutionary trajectories of Mycobacterium avium subsp. paratuberculosis in US dairy cattle using dimensionality reduction techniques

Dr. Scott Wells[1], Dr. Yuanyuan Wang[2]
[1]University Of Minnesota, St. Paul, USA, [2]Centers for Disease Control and Prevention, Atlanta, USA

Pathogenomics Genotyping and Map Diversity - AM Session (1), Main Auditorium, Printworks, June 13, 2022, 9:15 AM - 10:30 AM

While a number of Map strains have been identified through comparative analysis of different ruminant hosts, differentiating within strains remains challenging. In this study, we provide an exploratory framework to examine the evolution of Map from time-series whole genome sequence (WGS) data sequenced from longitudinal samples in Minnesota, New York, Pennsylvania, and Vermont to address the challenge of differentiating between samples with limited number of homogenous variants and large number of heterogenous variants.

We performed analyses on two levels: single nucleotide variants and population structures. To extract patterns from variants, we implemented Temporal Non-negative Matrix Factorization (NMF), which deconvolutes input genotype data into two low-rank non-negative matrices at each time point and optimizes both between matrices and across time points. Temporal NMF accounts for temporal variations in the pathogen genomes to obtain an evolving panel of template strains in which mixture samples can be identified. To inspect the population structure of Map, we performed two qualitative comparative analyses using an array of input data including allele frequencies, strain abundances, and Hamming distances, in combination with two dimension reduction methods to visualize the population structure.

Our results demonstrated that dimension reduction techniques can effectively reduce the noise from outliers to provide significantly better resolution in both within-herd and between-herd population structures than the baseline scenarios where dimension reduction was not performed. This novel computational method derived an evolving panel of 15 template strains from 525 whole genome sequences of Map, in which high proportions of heterogenous variants were observed. This work is the first attempt to account for temporal variations among WGS data in mycobacterial infections. These comparative analyses represented the first attempt to use dimension reduction techniques prior to constructing phylogenies for Map, with potential application to other pathogens.

Involvement of M. avium ssp. paratuberculosis specific genomic regions LSP14/15 in zinc homeostasis

Dr. Elke Goethe[1], Kristin Laarmann[1], Janita Lührs[1], Ayla Gieseke, **Prof Ralph Goethe[1]**
[1]Tierärztliche Hochschule Hannover, Hannover, Germany

Pathogenomics Genotyping and Map Diversity - AM Session (1), Main Auditorium, Printworks, June 13, 2022, 9:15 AM - 10:30 AM

Maintenance of zinc homeostasis is essential for bacterial survival. In the pregenomic era, we identified a 38 kb pathogenicity island in M. avium ssp. paratuberculosis (MAP), which was later shown to be part of the MAP-specific genomic regions LSP14/15. LSP14/15 harbour several predicted metal uptake systems. Due to the presence of binding motifs for the ferric uptake regulator Fur, these systems were initially assumed to support iron homeostasis. In recent years, we could show that these genetic elements were acquired to support MAP zinc homeostasis: Transcriptome analyses of MAP zinc starvation cultures revealed the zinc dependent regulation of 45 genes, 35 of which are presumably regulated by the global zinc uptake regulator Zur. In addition to the widely distributed zinc importer ZnuABC (map_0489-87c) we found two MAP-specific transporters MptABC (map_3736-34c) and ZnuABC-like (map3776-74c) on LSP14/15. We could show that all transporters clearly contributed to zinc uptake in a ZnuABC-deficient M. smegmatis (MSMEG) mutant.
Moreover, by promoter analyses in a MSMEGΔzur mutant we found, that znuABC and mptABC gene expression was regulated by Zur. We also investigated the phagosomal zinc status of MAP-infected macrophages treated with the zinc-specific fluorophore FluoZin3AM by immunofluorescence microscopy. No co-localisation of zinc and MAPCherry could be detected. qRT-PCR experiments of MAP derived from macrophages revealed a significantly induced mptA expression after 24h infection. These findings indicate that MAP in macrophages is subjected to intraphagosomal zinc deprivation. Finally, in infection experiments MAPwt exhibited significantly better survival in macrophages cultured in the presence of zinc than a MAPΔmptA mutant, which lacks a functional zinc uptake system. Overall, we provide strong evidence, that LSP14/15 are important for MAP zinc homeostasis and that ZnuABC, ZnuABC-like and MptABC must be considered as true zinc transporters. This might be of importance in the gut for colonisation and survival in the mucosal environment.

191

Genetic features of Mycobacterium avium subsp. paratuberculosis strains circulating in the West of France deciphered by Whole-Genome Sequencing

Mr Cyril Conde[1], Mr Thierry Cochard[1], Mr Maxime Branger[1], Ms Marie-Noëlle Rossignol[2], Pr. Christine Fourichon[3], Pr. Arnaud Delafosse[4], Ms Aurore Davergne[5], Dr. Alain Joly[6], Dr. David Ngwa-Mbot[7], Dr. Laurent Journaux[8], Pr. Raphael Guatteo[3], Dr. Laurent Schibler[9], Dr. John Bannantine[10], Dr. Franck Biet[1]

[1]ISP, INRAE, Université de Tours, Nouzilly, France, [2]INRAE, GABI, Université Paris-Sacley, Jouy-en-Jossas, France, [3]INRAE, UMR 1300 BIOEPAR, Nantes, France, [4]Groupement de Défense Sanitaire Orne, Alençon, France, [5]GDMA, Bois Guillaume, France, [6]Groupement de Défense Sanitaire Bretagne, Vannes, France, [7]GDS France, Paris, France, [8]Idele, Paris, France, [9]Allice, Paris, France, [10]USDA-Agricultural Research Service, Ames, United States of America

Pathogenomics Genotyping and Map Diversity - AM Session (1), Main Auditorium, Printworks, June 13, 2022, 9:15 AM - 10:30 AM

Paratuberculosis is a chronic infection of the intestine, mainly the ileum, caused by Mycobacterium avium subsp. paratuberculosis (Map) in ruminants. This enzootic disease is present worldwide and has a strong impact on the dairy cattle industry.
For this species, the typing tools do not make it possible to track the spread and microevolution of the strains. These limitations can be overcome by the application of Whole Genome Sequencing (WGS), particularly for clonal populations such as Map. WGS analyses can provide comprehensive genetic information, including information on genome evolution and discrimination of closely related strains.
The purpose of the present study was to undertake a whole-genome analysis of Map strains to identify accurate phylogenetic relationships between isolates and establish correlations between genomic traits and epidemiological data within a population of well documented-strains.

A set of 200 animal field strains, representative of the French Map population circulating in the West of France, were isolated from bovine of breed Prim'Holstein or Normande naturally infected by Map. For each strain isolated, all information about the animal is available including: herd prevalence, locations, serological status and excretion level. Map strains were sequenced on an Illumina MiSeq. Genomic sequences were assessed for potential contamination. Reads were aligned to a local reference to infer a SNP-based phylogeny.

This study provided 200 new genomes of French strain isolates from naturally infected animals. Pangenome analysis of this panel confirmed the degree of Map clonality.
SNP analysis provided accurate phylogeny able to distinguish each strain divided into 3 clusters independently of the cattle's breed. Interestingly, clusters seem associated with the two major MLVA profiles. A phylogeny was inferred with French Map isolates and with other Map isolates found across the world.
Relationships between genetic traits and epidemiological data will be investigated to better understand the transmission dynamics of the disease.

178

Phylogenetic analysis of Mycobacterium avium subsp. paratuberculosis in Australia

Mrs Rachel Hodgeman[1,2], Dr Rachel Mann[1], Mr Keith Savin[1], Mr Noel Djitro[1,2], Dr Simone Rochfort[1,2], Professor Brendan Rodoni[1,2]

[1]*Department Of Jobs, Precincts And Regions, Bundoora, Australia,* [2]*Latrobe University, Bundoora, Australia*

Pathogenomics Genotyping and Map Diversity - AM Session (2), Main Auditorium, Printworks, June 13, 2022, 11:00 AM - 12:00 PM

On behalf of Animal Health Australia, Agriculture Victoria has been the responsible authority for the Australian Johne's Disease Reference Laboratory for the last 18 years. The reference laboratory has the largest collection of Mycobacterium avium subsp. paratuberculosis (MAP) isolates in Australia, holding 291 isolates collected over the last 30 years. Here we report genome sequencing of 231 representative isolates from the Australian MAP collection. To determine where Australian isolates are positioned in a global context, and to assess the reliability of typing methods, phylogenetic analysis based on whole genome single nucleotide polymorphism (SNP) profiling, IS1311 genotyping, LSP gene analysis and average nucleotide identity were conducted on all isolates. Phylogenetic analysis of SNPs identified in the MAP core genome revealed 8 distinct clades within the Type C strains and 5 distinct clades within the Type S strains of MAP. There were fewer than 20 SNP differences across the core genome among some adjacent clades highlighting the monomorphic nature of MAP and suggesting that isolates from within a clade have risen from a common ancestor. Australian sheep strains clustered most closely to a sheep strain from Scotland (MAPMRIO103) and the Australian Bison strains clustered most closely to US bison type strains. No distinct phylogeographic clustering of MAP was observed in this study. IS1311 PCR and Restriction Enzyme Analysis (REA) intermittently generated incorrect results when compared to Long Sequence Polymorphism (LSP) analysis, whole genome SNP-based phylogenetic analysis, IS1311 sequence alignment and average nucleotide identity (ANI). These alternative methods generated consistent Map typing results. A published SNP based assay for genotyping Map was found to be unsuitable for differentiating between Australian and international strain types of Map.

221

Understanding the role of Mycobacterium avium subsp. paratuberculosis pathogenicity islands by generating gene knockouts using ORBIT

Shannon Duffy[1,2,3], Dr. Marcel Behr[1,2,3,4]

[1]Department of Microbiology and Immunology, McGill University Health Centre, Montreal, Canada, [2]McGill International TB Centre, McGill University, Montreal, Canada, [3]Infectious Diseases and Global Health Program, McGill University Health Centre, Montreal, Canada, [4]Department of Medicine, McGill University, Montreal, Canada

Pathogenomics Genotyping and Map Diversity - AM Session (2), Main Auditorium, Printworks, June 13, 2022, 11:00 AM - 12:00 PM

Background: Mycobacterium avium subsp. paratuberculosis (MAP)-specific genes are found within six genomic islands called large sequence polymorphisms of paratuberculosis (LSPps) which are predicted to make up MAP-associated pathogenicity islands. However, the roles of many of these genes are poorly understood. The ability to efficiently generate knockouts of MAP-specific genes would generate greater insight into their role in MAP pathogenesis.

Methods: Oligo-mediated Recombineering followed by Bxb1 Integrase Targeting (ORBIT) was validated for use with MAP. This technique was employed to target genes found within LSPp12 (MAP2189-94) and LSPPp15 (MAP3776-3). The role of these two operons were evaluated in vitro by measuring growth in defined media. C57BL/6 mice were intraperitoneally infected with ΔMAP2189-94 and ΔMAP3776-3 to assess their viability in vivo.

Results: No differences in growth were found between wild type MAP, ΔMAP3776-3, and ΔMAP2189-94 when grown in Middlebrook 7H9 media supplemented with mycobactin. ΔMAP3776-3 had reduced growth compared to wild type MAP when grown on 7H10 agar supplemented with 0.1% ferric ammonium citrate (FAC). But no differences in growth were observed when 7H10 agar was supplemented with 1% FAC. Mice infected with ΔMAP2189-94 and ΔMAP3776-3 had significantly lighter spleen weights and significantly less colony forming units (CFUs) in the spleen and liver compared to wild type at 4 weeks post-infection.

Conclusions: The introduction of the ORBIT system in MAP affords advantages in efficiency and labor-intensity over previously applied methods. These findings support previous work which suggest MAP3776-3 acts as an importer of iron alternative to mycobactin. Additional work is required to understand the role of MAP2189-94 in vitro. The loss of both of these MAP-specific genes decreased bacterial viability in mice, indicating they play a role in MAP survival. The implementation of the ORBIT system may assist future investigations seeking to understand MAP pathogenesis.

182

85 complex proteins as an important adhesion factor of Mycobacterium avium subsp. paratuberculosis in bovine sperm

Ms Jessica Caldeira[1], PhD Maria Aparecida Scatamburlo Moreira[1], Ms Edgar Diaz-Miranda[1], PhD José Guimarães[1], PhD Yung-Fu Chang[2], Ms Tiago Zilch[2], PhD Wagner Barbosa[1], PhD Abelardo Silva Junior[1], Ms Junnia Pena[1]

[1]Universidade Federal de Viçosa, Viçosa, Brazil, [2]Cornell University, Ithaca, USA

Pathogenomics Genotyping and Map Diversity - AM Session (2), Main Auditorium, Printworks, June 13, 2022, 11:00 AM - 12:00 PM

Mycobacterium avium subsp. paratuberculosis (MAP) has the 85 antigen complex, consisting of three proteins 85A, 85B and 85C. Studies have shown that these proteins are responsible for MAP adherence to host cells, where they bind to the fibronectin of these cells, constituting a critical factor in host colonization. Little is known about the interaction of MAP with sperm or which factors may be involved in this interaction. This study aimed to evaluate the interaction of MAP and bovine sperm in the presence of two complex 85 proteins. The 85A and 85B protein clones were expressed, purified and the hyperimmunization of six rabbits (three with the 85A protein and three with the 85B) was performed to obtain the antibodies. Two Nellore bulls were used as semen donors and were negative for MAP. Three situations were evaluated (i) sperm with MAP at a concentration of 10^6 cfu/ml (ii) sperm with MAP at a concentration of 10^6 cfu/ml in the presence of Ac 85A and (iii) sperm with MAP at a concentration of 10^6 cfu/ml in the presence of Ab 85B. A negative control with semen without MAP was used. The samples were evaluated for motility (0-100%) and spermatic vigor (0-5) under optical microscopy at different times. The interaction between MAP and sperm in the presence of Ac 85A and 85B was also visualized by scanning microscopy. The results showed that compared to the control sperm motility was significantly affected by the presence of MAP, not occurring when antibodies 85A and 85B were present. It was observed that MAP adhesion occurs in the intermediary part of the sperm and that this interaction was made difficult in the presence of 85A and 85B antibodies. Our study suggests that 85A and 85B proteins has an important role in the adhesion of MAP and bovine sperm.

302

Changes in the Fecal Microbiota of Calves Experimentally Infected with Mycobacterium avium subsp. paratuberculosis.

Dr. Marie-Eve Fecteau[1], Dr. Daniel Beiting[1], Dr. Raymond Sweeney[1]
[1]*University Of Pennsylvania, Kennett Square , United States*

Pathogenomics Genotyping and Map Diversity - AM Session (2), Main Auditorium, Printworks, June 13, 2022, 11:00 AM - 12:00 PM

Johne's disease (JD) is a widespread chronic gastrointestinal disease of cattle caused by Mycobacterium avium subsp. paratuberculosis (MAP). The mammalian gastrointestinal tract is colonized by hundreds of different species of bacteria, living in a delicate balance in a symbiotic relationship with the host. Dysbiosis can result in inflammation of the intestines and has been reported in cows naturally infected with MAP. It is however unknown if the dysbiosis observed in MAP-infected cattle is a consequence of MAP infection or a preexisting condition.

The purpose of this study was to investigate the impact of early MAP infection on the fecal bacterial populations in calves prior to, and following experimental infection, when compared to age-matched uninfected calves, using metagenomic analysis. Sixteen 4-week-old Holstein calves were used in this study. Pre-infection fecal samples were collected on all calves on Study Day 0. Twelve calves were orally challenged with 5 x 10^9 CFU MAP on Study Days 1 and 2. The remaining 4 calves were used as negative controls and housed separately from the infected calves. At 16 weeks post-infection, fecal samples were collected again on all 16 calves. All fecal samples were kept frozen at -70°C until analysis. Fecal samples were processed for DNA extraction, purification, and amplification using specific primers. Genomic sequencing was performed using NextSeq illumina platform.

All calves had similar fecal microbiota prior to infection, there was a statistically significant difference in microbial diversity between infected and uninfected calves at 16 weeks post-infection. The results of this study suggest that the dysbiosis observed in MAP-infected cattle is a consequence of MAP infection, and this dysbiosis may play a role in the intestinal inflammation observed in cattle infected with MAP.

287

Bio-load and Bio-type profile of Mycobacterium avium subspecies paratuberculosis isolates recovered from sheep and goats located in Himalayan region

Dr Adil Majid Bhat[1,6], Dr Hamidullah Malik[1], Dr Abdul Qayoom Mir[2], Dr Tufail Hussain[3], Dr Kundan Kumar Chaubey[4,5], Dr Saurabh Gupta[4,5], Dr Shoor Vir Singh[4,5]

[1]Division of Clinical Veterinary Medicine, Ethics and Jurisprudence, Faculty of Veterinary Sciences and Animal Husbandry, Shuhama, Srinagar, Srinagar, India, [2]Mountain Research Center for Sheep and Goats, Faculty of Veterinary Sciences and Animal Husbandry, Shuhama, Srinagar, Srinagar, India, [3]Department of Sheep Husbandry, Govt. of Jammu and Kashmir, Srinagar, India, [4]Department of Biotechnology, GLA University, Mathura, Mathura, India, [5]Microbiology Lab, Animal Health Division, Central Institute for Research on Goats, Mathura, Mathura, India, [6]Department of Veterinary Medicine (Directorate of Livestock Farms), Guru Angad Dev Veterinary and Animal Sciences University, Ludhiana, Punjab, Ludhiana, India

Pathogenomics, Genotyping and Map diversity - Poster Pitches, Main Auditorium, Printworks, June 13, 2022, 12:00 PM - 12:15 PM

Background: Small ruminants husbandry play important role in socio-economic life of humans in temperate regions of Kashmir valley. Johne's disease (JD) caused by Mycobacterium avium subspecies paratuberculosis (MAP), has been reported to be endemic in the small ruminant population of the country (Singh et al., (2014) - 23.3%), except temperate region. Present study investigated bio-load and biotype profile of MAP in the small ruminant population of Ganderbal district of valley.

Materials: A total of 288 (260 sheep; 28 goats) faeces were collected from 25 farmer's farms and 67 were from 4 government farms. Faecal samples were subjected to Ziehl-Neelsen (acid fast) staining. MAP isolates recovered using Herrold's Egg Yolk Medium with Mycobactin. Typing (IS900 PCR) and bio-typing (IS1311 PCR-REA) was carried out as per Sevilla et al. (2005).

Results: Prevalence of animals shedding acid fast bacilli indistinguishable to MAP was 32.1% in farmer's flocks and was 37.7% in 4 organized farms. Of 100 faecal samples positive in ZN staining were screened by IS900 PCR and 17 samples (5 goats; 12 sheep) were positive for MAP infection. Faecal samples positive (17) in IS900 PCR were biotyped (IS1311 PCR_REA) and 5 (4 sheep and 1 goat) were Type B (Bison). Necropsy of a sheep revealed infection of intestinal tissues with pigmented strain of MAP (yellow colouration of mucosal surface with prominent corrugations). IS1311 PCR-REA of DNA from tissue samples also revealed that MAP strain infecting sheep were 'bison type'.

Conclusion: This is the first evidence of existence of pigmented strain of MAP ' bison type' biotype. Study concluded that prevalence of MAP was high in sheep and goats reared in the temperate climate of Himalayas in country. 'Bison type' biotype was infecting sheep and goats and also reports existence of pigmented strains of MAP 'bison type' biotype.

190

Whole-Genome Analysis of Mycobacterium avium subsp. paratuberculosis IS900 distribution reveals strain type-specific modalities

Mr Cyril Conde[1], Dr. Marian Price-Carter[2], Mr Thierry Cochard[1], Mr Maxime Branger[1], Dr. Karen Stevenson[3], Pr. Richard Whittington[4], Dr. John Bannantine[5], Dr. Franck Biet[1]

[1]ISP, INRAE, Université de Tours, Nouzilly, France, [2]AgResearch Ltd., Hopkirk Research Institute, Palmerston North, New Zealand, [3]Moredun Research Institute, Midlothian EH26 0PZ, United Kingdom, [4]Faculty of Science, Sydney School of Veterinary Science, The University of Sydney, Camden, Australia, [5]USDA-Agricultural Research Service, Ames, United States of America

Pathogenomics, Genotyping and Map diversity - Poster Pitches, Main Auditorium, Printworks, June 13, 2022, 12:00 PM - 12:15 PM

Mycobacterium avium subsp. paratuberculosis (Map) is the etiological agent of paratuberculosis or Johne's disease that causes chronic intestinal inflammation in ruminants. The IS900 insertion sequence, specific for Map, has been used widely as an epidemiological marker and target for qPCR diagnosis. Recently, the complete genomes of strains belonging to the major phylogenetic lineages of Map have been made available. This has allowed us to investigate the distribution of IS900 in this slow growing bacterium. The objective of this study is to characterize the distribution of the IS900 element and how it affects genomic evolution and gene function of Map. A secondary goal is to develop automated in silico restriction fragment length polymorphism (RFLP) analysis using IS900.

Complete genomes from C-type and S-type (including subtypes I and III) lineages were chosen to represent the genetic diversity of Map. Computer analysis included software located IS900 using BLAST and determined fragments from complete genome (FASTA) developed with Biopython. Digital representation was provided using matplotlib. Profile comparisons were carried out using Bionumerics software. Upstream and downstream genomic regions flanking each IS900 copy were extracted and used to identify orthologous insertion site across genomes.

The program developed in this study allowed automated location of IS900 sequences. Between 16 to 22 copies of the IS900 sequence were found in the genomes studied. Nine IS900 insertion site locations were conserved across all genomes studied while smaller subsets were unique to a particular lineage. An in silico IS900 RFLP analysis was developed and profiles were compared by digital visualization.
This study provided a program making it possible to automate IS900 distribution analysis in Map genomes to enrich our knowledge on the dynamics of distribution of this IS for epidemiological purposes, for understanding the evolution within the Map species and studying the biological implication of the presence of IS900.

247

Deletion in mce4 gene exacerbates virulence of Mycobacterium avium subsp. paratuberculosis.

Dra. Maria Alejandra Colombatti Olivieri[1], **Dr Roberto Damian Moyano[1]**, Dr. Ariel Nagel[1], Dra. Maria Isabel Romano[1], Dra. Natalia Alonso[1]

[1]*Instituto de Agrobiotecnología y Biología Molecular INTA-CONICET, Hurlingham, Argentina*

Pathogenomics, Genotyping and Map diversity - Poster Pitches, Main Auditorium, Printworks, June 13, 2022, 12:00 PM - 12:15 PM

Mycobacterium avium subsp. paratuberculosis (MAP) is the causative agent of Paratuberculosis in ruminants, a chronic granulomatous enteritis. It is crucial to deepen the study of virulence genes involved in MAP pathogenesis.

The mce4 operon is involved in the uptake and utilization of cholesterol from host cells during infection, necessary for the successful cellular invasion and persistence in Mycobacterium tuberculosis. In a previous study, we demonstrated that lprG-p55 operon is involved in the virulence of K-10 strain. In the present study we evaluated the effect of mce4 and lprG-p55 deletion on the virulence of a local strain 6611.

To evaluate its virulence, we carried out infection assays in bovine monocyte-derived macrophages (BMDM) and mice model. For both assay we used the wildtype 6611 strain (6611-WT), the mce4 mutant (6611-Δmce4), and lprG-p55 mutant (6611-ΔlprG-p55). Initial screening in BMDM showed that mce4 deletion increased the survival of this strain in macrophages, compared to the 6611-WT strain at 1, 2 and 4 days post infection (dpi). On the other hand, ΔlprG-p55 was the strain with the lowest percentage of survival in macrophages at all evaluated times. Subsequently, we characterized the virulence of the mutant strains in BALB/c mice. All the strains were able to colonize mice spleens, with an average of 5x105 CFU/mL at 3 dpi. The ΔlprG-p55 strain had a significant decrease in the spleen CFU count from 20 dpi, compared to 6611-WT and Δmce4. In contrast, the Δmce4 mutant showed a significantly higher CFU count than 6611-WT and ΔlprG-p55 strains at 60 dpi.

The results obtained suggest that the deletion of the lprG-p55 operon leads to the attenuation of MAP while the deletion of the mce4 operon leads to higher virulence of the strain in both evaluated models. More studies are necessary to understand the role of these genes in MAP pathogenesis.

222

A streptomycin pre-treatment mouse model of Mycobacterium avium subsp. paratuberculosis infection

Shannon Duffy[1,2,3], Dr. Andréanne Lupien[2,3], Fiona McIntosh[2,3], Dr. Marcel Behr[1,2,3,4]
[1]Department of Microbiology and Immunology, McGill University Health Centre, Montreal, Canada, [2]McGill International TB Centre, McGill University, Montreal, Canada, [3]Infectious Diseases and Global Health Program, McGill University Health Centre, Montreal, Canada, [4]Department of Medicine, McGill University, Montreal, Canada

Pathogenomics, Genotyping and Map diversity - Poster Pitches, Main Auditorium, Printworks, June 13, 2022, 12:00 PM - 12:15 PM

Background: An experimental small animal model of Mycobacterium avium subsp. paratuberculosis (MAP) infection can help mechanistically determine the role of host and pathogen factors associated with disease. However, current mouse models of MAP infection either rely on a chemical treatment to cause inflammation or use a non-physiologic route of infection. An established model of infection with the enteropathogen Salmonella Typhimurium involves pre-treatment of streptomycin to reduce colonization resistance followed by oral infection with a streptomycin-resistant strain.

Methods: Oligo-mediated recombineering was employed to introduce a K43R mutation in the rpsL gene of MAP K10 to confer resistance to streptomycin. C57BL/6 mice were pre-treated with streptomycin prior to infection with 2 doses of 10^9 colony-forming units (CFUs) of streptomycin-resistant MAP. Infection outcomes were assessed by organ plating and histopathology. MAP infection was compared with Mycobacterium avium subsp. hominissuis (MAH), also made streptomycin-resistant, as a control for environmental M. avium.

Results: Pre-treatment with streptomycin prior to MAP infection led to a 2-log increase in MAP CFUs in the colon compared to no pre-treatment. This infection was sustained at 3 months post-infection. MAP colonization was also established in the mesenteric lymph nodes of mice. This model did not lead to systemic infection in the spleen or liver. Infection with MAH resulted in a transient infection in the same experimental model that was cleared by 3 months.

Conclusions: A streptomycin pre-treatment MAP infection model in mice may enable mechanistic investigation of host and microbial factors associated with infection and disease. This model may also act as a platform for assessing novel MAP vaccines and therapeutics.

323

Invited Speaker for Control Programs and Education

Dr Herman Barkema

Control Programs and Education - Invited Speaker, Main Auditorium, Printworks, June 13, 2022, 1:30 PM - 2:15 PM

Five learning points

1. Calf-to-calf transmission has been ignored in Mycobacterium avium subsp. paratuberculosis (MAP) control, potentially contributing to a lack of success in most control Johne's disease (JD) programs .

2. The genetic basis of susceptibility to MAP infection needs to be sufficiently characterized.

3. Although new MAP tests are available, they must be well validated well before being used in the field.

4. Good communication between veterinarian and farmer is essential when controlling JD.

5. The motivation to adopt JD control programs should not be limited to the potential association with Crohn's disease, as the economic losses JD causes to the dairy industry justify its control.

282

Part 3: UK approach to Johne's Disease control -Top tips for effective control

Mr Pete Orpin[1,2], Mr Dick Sibley[2,3]
[1]Myhealthyherd, Cosby, United Kingdom, [2]Special Lecturer Nottingham University, Nottingham, United Kingdom, [3]Westridge Veterinary Group, Tiverton, United Kingdom

Control Programs and Education - PM Session (1), Main Auditorium, Printworks, June 13, 2022, 2:15 PM - 3:30 PM

Introduction: The UK JD program has been driven by the use of milk ELISA testing and effective risk management. An estimated 71% of herds utilise improved farm management and strategic testing, culling and management. Effective segregation of high-risk cows and culling decisions are challenging areas to manage throughout the year.

Methodology: To improve farmer engagement with controls most recent veterinary education has focused on the targeted application of JD control to high risk places or time periods.
Focus has centred on the creation of a low risk maternity and calf pen areas ("Green calving and calf line") to minimise disease transmission.
Categorisation of test results to aid decisions to cull immediately, end of lactation, do not breed or breed to terminal sire has been encouraged.
Utilising red ear tags to clearly identify high risk animals preventing them entering the 'green calving line" is a clearly understood concept by farmers.
Identifying times of the year where the "green calving line" can be most simply applied and the use of sexed semen has allowed for sourcing of low risk replacements from low risk environments greatly simplifying controls.
Calves from test positive cows are tagged red and not allowed to enter "green calf areas" to minimise calf to calf transmission.

Discussion: A robust JD control plan can only be delivered by changes in farmer behaviour.
Utilising simple language and developing readily grasped concepts based on solutions rather than problems is central to JD control. Focusing the mind on creating ways of protecting 30% of the calf crop destined for herd replacements is easier to achieve than delivering robust control for 100% of the year. Protecting the maternity environment from contamination combined with cost effective breeding and culling decisions is vital for cost effective JD control.

231

Dynamics of MAP infections in small structured cattle farms

Dr Johannes Lorenz Khol[1], Thorben Liening-Ewert[2], Laura Auböck[1], Dr. Paul Ortner[2], Dr. Josef Kössler[2], Dr. Alexander Tichy[3], Dr. Michael Dünser[4], Dr. Christian Mader[5]

[1]University Clinic for Ruminants, University of Veterinary Medicine Vienna, Vienna, Austria, [2]Regional Veterinary Office Tyrol, Innsbruck, Austria, [3]Bioinformatics and Biostatistics Platform, University of Veterinary Medicine Vienna, , Austria, [4]Austrian Agency for Health and Food Safety, Linz, Austria, [5]Veterinary Health Service Tyrol, , Austria

Control Programs and Education - PM Session (1), Main Auditorium, Printworks, June 13, 2022, 2:15 PM - 3:30 PM

The agricultural system in the Austrian province of Tyrol is characterized by small, family owned cattle farms and alpine pasture, combining animals from different farms, during summer. In 2013, a voluntary survey and control program for MAP in cattle was established, based on the evaluation of boot swab samples. About 4,600 farms, representing 70% of the Tyrolean dairy cattle, are participating in this program.
In course of the program, MAP-positive farms are detected by boot swab sampling, followed by single animal testing (fecal culture and/or serum ELISA) and removing of positive animals, as well as establishing basic hygienic measures. Thereby, the initial prevalence of 7.5% positive herds in 2013 decreased to 0,5% in 2019.
Longitudinal investigation of individual animal results showed, that MAP-shedding within a herd decreased markedly after removal of single positive animals. Additionally, many cattle showed negative individual results and farms stayed MAP-negative in consecutive boot swab samplings thereafter. This indicates possible passive shedding in some animals. Furthermore, fade out of the disease, after removal of MAP-shedding animals and introduction of basic hygienic measures seems to be achievable.
Analysis of risk factors showed that, the use of common alpine pasture seems not to be a significant contribution to the transmission of MAP (OR: 1.194, 95% CI: 0.821/2.311). The same was found for sharing of equipment (OR:0.997, 95% CI:0.806/1.233). On the other hand, rearing of calves with milk replacer, instead of whole milk, significantly decreased the chance to be MAP positive (OR: 0.458. 95% CI: 0.233/0.900).
The results of the Tyrolean program for the survey and control of paratuberculosis indicate, that the dynamics of MAP infections in small structured cattle farms may differ from other agricultural systems. This might contribute to the successful reduction of the MAP-herd prevalence by removing of single positive animals, combined with basic hygienic measures.

317

Predicting positive ELISA results in dairy herds with a preferred herd status in a milk quality assurance programme for paratuberculosis

Dr. Maarten F. Weber[1,2], Dr. Marian Aalberts[1], Dr. Thomas Dijkstra[1], Prof. dr. Ynte H. Schukken[1,2,3]
[1]Royal GD, Deventer, the Netherlands, [2]Dept. of Population Health Sciences, Faculty of Veterinary Medicine, Utrecht University, , the Netherlands, [3]Quantitative Veterinary Epidemiology, Dept. of Animal Sciences, Wageningen University, , the Netherlands

Control Programs and Education - PM Session (1), Main Auditorium, Printworks, June 13, 2022, 2:15 PM - 3:30 PM

Herds participating in the Dutch milk quality assurance programme for paratuberculosis are assigned a herd status based on results of herd examinations by ELISA on individual serum or milk samples. Confirmation of positive (S/P≥1.0) ELISA results by individual faecal PCR assay is optional. Test-negative herds are assigned status A; surveillance of these herds consists of biennial herd examinations.

To control any undetected Map infections in status A herds, preventive management measures are indicated. However, farmers falsely believing that their status A herds are Map-free may refrain from preventive measures. Therefore, the aim of the present study was to develop a predictive model to alert farmers with status A if they are at increased risk of positive ELISA results.

Using data of 8,566 dairy herds with status A and ≥50 adult cattle in January 2016 as well as ≥50 ELISA results in 2014-2015, a logistic regression model was built with the probability of positive ELISA results in January 2017 – June 2019 as dependent variable and province, soil type, herd size, proportion of cattle born elsewhere, time since previous positive results and the 95%-percentile of S/P ratios in 2015-2016 as explanatory variables.

As internal validation, the model was applied to predict positive ELISA results in January 2019 – June 2021 in 8,026 herds with status A in January 2019. The model had an ROC curve with an AUC = 0.76 (95% CI: 0.75, 0.77). At a cut off predicted probability π=0.40, 25% of status A herds would receive an alert with positive and negative predictive values of 0.53 and 0.83, respectively.

The results of this study indicate that prediction of the risk of future positive ELISA results in status A herds is feasible. This might stimulate farmers with an increased risk of spread of Map to take appropriate control measures.

291

The Irish Johne's Control Programme: a qualitative approach to understanding the motivations for farmer and veterinary practitioner engagement and providing information effectively through co-design to improve recruitment, adherence, and retention.

Mrs Alison Burrell[1], Dr Margaret M Good[3], Associate Prof Conor G McAloon[2], Dr David A Graham[1], Prof Simon J More[2], Mr Lawrence Gavey[1]

[1]Animal Health Ireland, Carrick on Shannon, Ireland, [2]School of Veterinary Medicine, University College Dublin, Belfield, Ireland, [3]Independent Researcher and Private Consultant, Dun Laoghaire, Ireland

Control Programs and Education - PM Session (1), Main Auditorium, Printworks, June 13, 2022, 2:15 PM - 3:30 PM

The IJCP is a voluntary Johne's disease control programme for cattle farmers and supporting Approved Veterinary Practitioners (AVP). It aims to provide farmers with management advice to prevent or reduce levels of JD through yearly whole herd testing and risk assessments. However, uptake and completion of these funded yearly requirements is lower than expected among participating farms. This results in incomplete assessment, missed testing reimbursement and higher resource burden for AVPs and the programme.

Research highlights the importance of considering socio-psychological factors that influence farmers' adoption of recommendations or enrolment in a voluntary programme. The COM-B model aims to understand the internal and external influences of behaviour in terms of an individual's capability, opportunity and motivation.

This 3-part study examines the barriers and facilitators to farmers' initial and continued engagement with the IJCP and its yearly requirements, highlight AVPs' experiences of supporting their clients, explore future needs of farmers within the IJCP and identify how best to communicate the complexities of the programme to end-users.

Method: Part 1 involves semi-structured qualitative interviews framed by COM-B with farmers participating in the IJCP. Using purposeful criterion sampling, farmer recruitment (n=20 approx.) is based on records of completion of IJCP requirements from 2020. Part 2 invites all AVPs (n=330) to complete an online qualitative questionnaire.

Results: Qualitative data will be available from early 2022 and will inform Part 3: a co-design workshop with participating farmers to design new ways of effectively communicating complex concepts and practices of the IJCP to end-users. This workshop will take a Design Thinking structure: Empathise, Define, Ideate, Prototype, Test.

Conclusion: Recruitment, adherence, and retention of farms to the IJCP will be improved with an understanding of the socio-psychological factors that influence farmers' behaviour. Further, co-design with end-users will allow more effective delivery of information.

280

Part 1: UK approach to JD control -Farmer Engagement using low cost surveillance, risk assessment and prevalence prediction

Mr Pete Orpin[1,2], Mr Dick Sibley[1,3]
[1]Myhealthyherd, Cosby, United Kingdom, [2]Special Lecturer Nottingham University, Nottingham, United Kingdom, [3]Westridge Veterinary Group, Tiverton, United Kingdom

Control Programs and Education - PM Session (1), Main Auditorium, Printworks, June 13, 2022, 2:15 PM - 3:30 PM

Introduction: Farmer engagement must be achieved for success with Johne's Disease (JD) control. In the UK this was achieved using education, risk assessments and targeted surveillance utilising 30 cow milk ELISA screens and prevalence prediction over a 5 year period before developing the National Johne's Management Plan.

Methodology: Milk processors were encouraged to host over a 80+ farmer events nationally to educate farmers on the importance of JD control. 1200 farmers engaged directly through milk processors and a further 1000 through a regional health scheme. Farmers were asked to submit milk samples from the 30 highest risk cows in the herd (disappointing milk yields, high cell counts, history mastitis/ lameness of symptoms of JD). Contemporary risk assessments identified risk of JD introduction and spread. A proportion were entered into Myhealthyherd, a web-based health planning tool, for traffic light scoring. High risk herds with red risks of entry and spread were prioritised for further surveillance if test negative. The program combined the risk score and test prevalence to estimate of future true herd prevalence. The process was further extended over 12 year period to engage the majority of UK farmers.

Results: The most important result was the creation of a social norm to test and manage for JD nationally. The targeted 30 cow screen and education was provided free by the milk processor. Data from 2502 herds revealed that 54% of dairy herds were high risk of disease entry with 77% high risk of within herd spread.

Discussion: Simple, cost effective methods of providing education, surveillance and risk management are crucial for JD engagement. The combined targeted 30 cow screen and risk assessment were easily implemented engaging the farmer and vet in next steps of JD control. Targeted milk ELISA testing provides a low cost method of surveillance.

244

Changes in Johne's Disease Control Practices in Ontario Canada between 2013 and 2019

Dr. Jamie Imada[1], Dr. Steven Roche[1,2], Dr. Cathy Bauman[1], Dr. David Kelton[1]
[1]University Of Guelph, Guelph, Canada, [2]ACER Consulting, Guelph, Canada

Control Programs and Education - PM Session (2), Main Auditorium, Printworks, June 13, 2022, 4:00 PM - 4:45 PM

Johne's disease (JD) control is often based on the culling of positive animals and the adoption of management practices that minimize exposure of calves to the pathogen Mycobacterium avium subspecies paratuberculosis. From 2010 to 2013, Ontario, Canada instituted a voluntary Johne's control program consisting of whole herd testing and a risk assessment and management plan (RAMP). The RAMP consisted of 38 questions which evaluated 5 different management areas to characterize herd risk for MAP introduction and within herd spread. The RAMP produced a numerical score, with higher scores associated with a higher risk. The RAMP focused on animal purchases, calving management, calf management, and heifer and cow cleanliness and management. In the summer of 2019, the RAMP was repeated on 180 farms that had both participated in the JD program of 2010-2013 and had bulk tank milk ELISA results from 2013 and 2017. Preliminary analysis indicates that many producers changed their management over the four to seven year period. Herds changed their cattle buying practices, with a marked reduction in purchasing from multiple sources and more herds refraining from buying in animals. However, overall scores were higher in the 2019 RAMPs than in 2013. Current RAMPs indicated that fewer farms were utilizing individual calving pens in 2019 than in 2013 (11% vs 32%), yet more farms had policies in place to deal with sick or suspect JD cows entering the maternity area (93% vs 80%). Management changes have occurred over time, some of these represent increased risk (crowded maternity pens) and others indicate decreased risk (closed herd, protocols in place for JD positive cows) for MAP introduction and transmission. These results highlight the importance of frequent risk assessments and the documentation of changes to management practices on farm to more accurately assess herd disease risk.

238

Raising awareness on paratuberculosis in Africa

Mr Julius Boniface Okuni[1], Mr Judah Ssekitoleko[1], Mr Kamal Eltom[2], Ms Wisal A. Elmagzoub[2], Ms Sanaa M. Idris[2], Mr ElSagad Eltayeb[3], Mr Ahmad Amanzada[4], Mr Joseph Omega[5], Ms Franziska Tanneberger[6], Ms Lonzy Ojok[1], Mr Ahmed Abd El Wahed[6]

[1]Makerere University, , Uganda, [2]University of Khartoum, Khartoum , Sudan, [3]Ibn Sina Specialised Hospital , Khartoum, Sudan, [4]University Medical Centre Goettingen, Goettingen, Germany, [5]University of Eldoret, Eldoret, Kenya, [6]Leipzig University , Leipzig , Germany

Control Programs and Education - PM Session (2), Main Auditorium, Printworks, June 13, 2022, 4:00 PM - 4:45 PM

Mycobacterium avium subsp. paratuberculosis (MAP) has been studied extensively in Europe and the United States of America. MAP occurrence, distribution, risk factors and economic impacts have not been fully investigated in Africa. Around 19 African countries have reported the diseases. The few published reports on MAP in sub Saharan Africa came mainly from Sudan, Uganda, Tanzania, Kenya and Ethiopia. However, this does not mean that other countries in the region are free from the disease. We have created a social media initiative to raise awareness about MAP as a neglected disease in Africa. The consortium has established a series of summer schools to train African young scientists on case identification and laboratory diagnostics. Two MAP molecular detection mobile laboratories have been deployed in Africa to enhance epidemiological studies. Moreover, workshops with decision makers and farmers have been conducted to identify research gaps and a simple web-based platform is being planned to allow community members to report suspected cases in domestic as well as wild animals. Future studies shall focus on creating a disease control strategy suiting the African settings. Basic research must aim at identifying resistance or tolerance among local breeds as well as the virulent MAP strains in Africa. Zoonotic potential of MAP in Africa is another research gap which needs to be addressed since the relationships between livestock and humans in Africa are very strong, as in some communities in Africa, humans share the dwelling and water sources with their livestock. Furthermore, most of the animal products are consumed untreated or under-cooked by humans. In conclusion, our activities helped in increasing the knowledge about the disease in Africa and would provoke international foundations and local stakeholders to allocate funds for more research on MAP in Africa and on establishing control programmes.

262

Different Mycobacterium avium subspecies paratuberculosis burden implies different aims of Johnes' control strategies in the three main dairy production regions of Germany

Dr. med.vet. habil. Karsten Donat[1], Mette Krieger[3], PD Dr. med. vet. habil. Amely Campe[3], PD Dr. med. vet. habil. Ingrid Lorenz[4], Dr. med. vet., PhD Susanne Eisenberg[5]

[1]Thuringian Animal Disease Fund, Victor-Goerttler-Strasse 4, Jena, Germany, [2]Clinic for Obstetrics, Gynecology and Andrology with Veterinary Ambulance, Justus-Liebig-University Giessen, Gießen, Germany, [3]University of Veterinary Medicine Hanover, Department of Biometry, Epidemiology and Information Processing, WHO Collaborating Centre for Research and Training for Health at the Human-Animal-Environment Interface, Hanover, Germany, [4]Bavarian Animal Health Service, Poing, Germany, [5]Animal Disease Fund of Lower Saxony, Hanover, Germany

Control Programs and Education - PM Session (2), Main Auditorium, Printworks, June 13, 2022, 4:00 PM - 4:45 PM

On-farm environmental sampling has been proven as an effective method for herd-level diagnosis of Mycobacterium avium subsp. paratuberculosis (MAP) infection and to estimate between-herd prevalence. Management factors and herd-size are known as important risk factors for herd prevalence. Within Germany, dairy herd sizes and management as well as breeds differ between regions implicating regional differences in MAP prevalence. The objective of this cross-sectional study was to assess the between-herd prevalence of paratuberculosis for different regions in a standardized approach.

For this study, 457 randomly selected dairy farms from three regions of Germany (North: 183, East: 170 South: 104) were sampled between 2017-2019. Enrolled farms were randomly selected from the German database on animal identification and registration (HI-Tier) and stratified according to herd size. Environmental samples (boot-swabs, aggregate faeces and/or liquid manure samples) were cultured and analysed using an IS900-qPCR for MAP determination.

Of the 457 selected farms, 94 had at least one MAP positive environmental sample with significant differences between regions regarding the apparent (North: 12.02%, East: 40.59%, South: 2.88%) or corrected true (North: 14.8%, East: 50.1%, South: 3.6%) between-herd prevalence. MAP was detected by faecal culture, qPCR or both methods in 86 out 450 boot-swabs, 38 out of 194 liquid manure samples, and 40 out of 269 aggregate faeces samples. In total, 121 samples were positively tested by faecal culture compared to 142 when qPCR was applied, with only moderate agreement between both methods.

In conclusion, the dairy production regions in Germany differ markedly in between-herd prevalence of paratuberculosis, indicating the need for control approaches with different aims. Different approaches are needed to deal with the disease striving for different aims ranging from on-farm prevalence reduction to mitigating the risk of between-herd transmission.

290

Completing herd investigations using a mobile App with links to real-time data

Ms Lorna Citer[1], **Lawrence Gavey[1]**, Mr Padraig O'Sullivan[2], Dr David Graham[1]
[1]Animal Health Ireland, Carrick on Shannon, Ireland, [2]Irish Cattle Breeding Federation, Bandon, Ireland

Control programmes and Education - Poster Pitches, Main Auditorium, Printworks, June 13, 2022, 4:45 PM - 5:00 PM

Herdowners registered with the Irish Johne's Control Programme (IJCP), must have a herd-level investigation completed by an Approved Veterinary Practitioner (AVP) when their herds are PCR-positive for MAP. The investigation is considered complementary to any previous risk assessment for Johne's disease that may have occurred.
All data collected through the IJCP is held in the Irish Cattle Breeding Federation's (ICBF) database which also contains genetic and production information for all registered cattle in Ireland.
Animal Health Ireland (AHI) commissioned the ICBF to develop a mobile application (mobile App) which enables the user to input data from herd investigations into the ICBF database in real-time. An additional feature is access to data about the dam's test history and livestock introductions for the previous 10 years at the time the investigation is carried out.
Members of the AHI Johne's Disease Technical Working Group developed an epidemiological questionnaire whose design ensured minimal use of free text to facilitate future data analyses whilst retaining technical robustness.
AVPs were surveyed to determine which mobile devices were commonly used in routine farm visits and this information informed the design of the Application.
The App is suitable for use with both Android and iPhone devices and is password protected to ensure compliance with the EU General Data Protection Regulation (GDPR).
The AVP only views herdowner information once the herdowner has registered for an investigation and the AVP's name had been linked to it. Data collected during the investigation is uploaded to the database and a completed copy of the questionnaire may be emailed to the herdowner from the mobile device, along with a series of recommendations for the management of high-risk animals.
Feedback from herdowners and AVPs has been positive, the quality of the data collected has improved and administrative time reduced for AVPs and AHI.

230

A preliminary analysis of vRAMP responses for dairy herds entering the voluntary Johne's disease control programme in Northern Ireland during its first year of operation

Miss Kayleigh Meek[1], Dr Sam Strain[2], Professor Niamh O'Connell[1], Professor Irene Grant[1]
[1]Queens University Belfast, Belfast, Northern Ireland, [2]Animal Health and Welfare Northern Ireland, Armagh, Northern Ireland

Control programmes and Education - Poster Pitches, Main Auditorium, Printworks, June 13, 2022, 4:45 PM - 5:00 PM

Northern Ireland (NI) has previously never implemented a province-wide Johne's Disease control programme for dairy herds. This has made it difficult to evaluate on-farm risks of infection spread between and within NI dairy farms. Animal Health and Welfare Northern Ireland have introduced a voluntary JD control programme with three key goals – bioexclusion, biocontainment and market reassurance. To meet these, herds are provided with recommendations to reduce JD risks, based upon an on-farm risk assessment and management plan (vRAMP) undertaken by approved veterinary practitioners. It is advised that herds undertake whole herd JD testing as part of the control programme. To date, 152 herds have enrolled onto the programme. The vRAMP results have been analysed using descriptive and non-parametric statistics in SPSS. This preliminary analysis showed that 24.3% of herds had previously completed a whole herd JD test, with 62.2% of these having undertaken individual milk testing. 33.6% of herds reported suspected cases of JD, where cattle showed clinical signs, and 12.5% of herds indicated a confirmed case within the past five years. Various risk factors were scored on a scale (1, 4, 7 and 10, indicating low to high risk) and compared against herds which reported confirmed cases. Non-parametric tests showed that four risk factors were significantly ($p<0.05$) more likely to occur in herds with a previously confirmed case, including utilising the maternity pen as a sick pen ($p = 0.019$), known JD positive cows using the same calving pen as JD negative cows ($p = 0.001$), calves suckling from multiple cows ($p = 0.025$) and weaned calves being exposed to manure or other cows ($p = 0.006$). This is the first time risk factors that may be important for JD transmission within NI dairy herds have been identified. This information will be used to refine future on-farm advice.

307

Irish Johne's Control Programme description and progress

Lawrence Gavey[1], Simon More[2], Lorna Citer[1], David Graham[1]
[1]Animal Health Ireland, Carrick On Shannon, Ireland, [2]Centre for Veterinary Epidemiology and Risk Analysis, UCD School of Veterinary Medicine, University College Dublin, Ireland

Control programmes and Education - Poster Pitches, Main Auditorium, Printworks, June 13, 2022, 4:45 PM - 5:00 PM

An update to the Irish Johne's Control Programme (IJCP) is presented.
The IJCP provides a long-term approach to the voluntary control of Johne's disease (JD), strongly supported by all actors in the Irish dairy industry and managed by Animal Health Ireland.
The IJCP has four objectives, consistent with the estimated herd prevalence of 30%: facilitate farmers' protection against spread of infection to uninfected farms, reduce the level of infection when present, assure markets of JD control in Ireland and improve calf health and farm biosecurity.
Key IJCP elements are annual veterinary risk assessment and management plans (VRAMPs), whole herd tests (WHT) by ELISA on blood or milk samples, ancillary faecal PCR testing of animals with positive or inconclusive ELISA results, and investigations of infected herds. There are pathways for assurance of test negative herds and for management of test-positive herds.
The costs of these activities are shared by key stakeholders. Milk processors support ELISA testing of dairy herds for WHTs, and the Irish government (DAFM) funds VRAMPs for dairy herds. Ancillary PCR testing and investigations of confirmed infection are fully funded for both beef and dairy herds.
The programme is supported by advice from a Technical Working Group; approval of veterinarians, tests and laboratories; documentation of policies and procedures; stakeholder leadership and direction; training; and broad communications and awareness activities.
Programme data is held on an accessible industry database where it is linked to animal and herd data on movements, Tb testing, and breeding; and from which herdowners and veterinarians can readily access results, reports and alerts.
Participation is voluntary. Registrations approached 2,000 herds by the end of 2021. National bulk tank milk surveillance identifies high-risk herds for recruitment.
Tools and systems are refined to address emerging issues and enhance the value of the programme.

196

Important Role for Abattoir Monitoring and Vaccination in Control Programs for Bovine Paratuberculosis Due to Mycobacterium avium subsp. paratuberculosis (Map)

Dr. Ian Links[1], Dr Heinrich Dahmen[2]
[1]Graham Centre For Agricultural Innovation, Charles Sturt University, Wagga Wagga NSW, Australia, [2]Tieraerztliche Praxis Dres. Dahmen & Duenner, Pruem, Germany

Control programmes and Education - Poster Pitches, Main Auditorium, Printworks, June 13, 2022, 4:45 PM - 5:00 PM

Background: Abattoir monitoring has proven highly sensitive, specific and cost effective in detection of paraTB in sheep in Australia. Surprisingly, there has been little international support for monitoring cattle at slaughter, despite mandated inspection for tuberculosis in most countries.

Dahmen at ICP14 (2018) reported detection of paraTB in 143 cattle in an abattoir in Germany while routinely inspecting for tuberculosis. On confirmatory histopathology the cases ranged from early to advanced pathology, demonstrating the practicality of screening cattle viscera by visual inspection.

Discussion: Despite increasing prevalence and economic loss to the cattle industry worldwide, a recent review confirmed formal control programs in only 22 of 48 countries. Surveillance deficits, leading to poor definition of herd and regional prevalence, were significantly impacting control programs. Concurrently, public health authorities were relying on these paraTB control programs to minimise the risk of Map entering the food chain.

In NSW Australia, abattoir monitoring filled this surveillance gap in sheep replacing costly on-farm testing. From 1999 to 2009 7.6million sheep carcases were examined from more than 32,000 sheep consignments (Links et al 2021). Feedback on sub-clinical disease prevalence to individual producers led to vaccination of more than 12 million sheep with Gudair® killed whole cell vaccine.

Regional prevalence fell from 2.4% in 2000 to 0.8% in 2009, accompanied by a progressive reduction in high prevalence consignments and absence of clinical disease, confirming declining on-farm prevalence.

Conclusion: Monitoring cattle in a German abattoir and sheep in Australia confirms the feasibility and benefits of routine monitoring of cattle for ParaTB at slaughter. Such a program should be implemented internationally as a routine adjunct to inspection for tuberculosis. Vaccination with killed whole cell vaccines in cattle has the potential to replicate the progress in disease control seen in sheep and mitigate the risk of Map entering the food chain.

253

How multiplex diagnostics can help in the rapid screening of Johne's Disease: A proposed surveillance method.

Dr Natasha Gordon[1], Dr Jimena Tejerina[1], Dr Dave Whittaker[1], Mr Howard Moore[1], Dr Richard Janeczko[1]
[1]*Pictor, Parnell, New Zealand*

Control programmes and Education - Poster Pitches, Main Auditorium, Printworks, June 13, 2022, 4:45 PM - 5:00 PM

Johne's disease, caused by Mycobacterium avium paratuberulosis (MAP), is present throughout much of the world. It results in chronic enteritis of cattle, leading to weight loss, a drop in milk production and reduced fertility. The economic losses in the diary industry due to infected animals amounts to millions of euro/dollars every year. A sensitive, high throughput assay for use in herd and individual animal surveillance is a much needed tool for this industry.

The PictArray™ Johne's test is a multiplex ELISA which targets multiple MAP antigens to provide a more thorough picture of Johne's disease compared to on market singleplex assays. Multiple biomarker targets ensure that different stages of the disease as well as different animal immune responses are captured within the multiplex test data, affording higher diagnostic sensitivity than assays utilising a single biomarker. Pictor proposes to introduce the PictArray™ Johne's test for whole herd testing using bulk tank milk samples. If a herd is identified as positive, individual animal testing can then proceed to confirm the infected animals. This assay will facilitate improved biosecurity and support trade as tests can be performed quickly and accurately when importing, exporting or moving livestock and associated products between local locations, resulting in safe transportation without the risk of spreading infection.

The PictArray™ Johne's test is a laboratory based assay, requiring a small volume of milk for testing. It is a colourimetric assay which is quantified on the PictImager CL2 reader (Scienion) and produces a report for each sample tested. Samples are characterised as positive, borderline or negative for each biomarker tested. A final combined assay result is also reported based on the analysis of each biomarker. The PictArray™ Johne's test will be a positive addition to national control programs to help in the eradication of Johne's disease from the diary industry.

246

Investigating the dynamics of Johne's disease in dairy herds after a province wide voluntary control program in Ontario Canada

Dr. Jamie Imada[1], Dr. Steven Roche[1,2], Dr. Cathy Bauman[1], Dr. David Kelton[1]
[1]University Of Guelph, Ontario, Canada, [2]ACER Consulting, Guelph, Canada

Control programmes and Education - Poster Pitches, Main Auditorium, Printworks, June 13, 2022, 4:45 PM - 5:00 PM

Current Johne's disease (JD) control is based on the culling of infected animals and the adoption of management practices that minimize exposure of young stock to the pathogen Mycobacterium avium subspecies paratuberculosis (MAP). From 2010 to 2013, Ontario, Canada instituted a voluntary Johne's control program in dairy cattle, consisting of whole herd cow-level testing and risk assessment. In addition, using a modified milk ELISA technique (IDEXX) with an optical density cut-off of 0.089, province wide bulk tank (BT) milk testing was used to estimate the prevalence of JD high risk herds at the end of the control program, and 4 years later in 2017 to assess changes in prevalence. The risk assessment used in the Ontario Johne's program, which was repeated on a subset of farms in 2019, had 38 questions focusing on 5 different management areas to characterize herd JD risk. Logistic regression models were built using risk assessment results from the original program and the 2019 follow up study, using BT results as the outcome variable. The final logistic models indicated that farms that reported having calving pens with more than one cow in it >25% of the time at the time of their original risk assessment were roughly 3 times more likely to have a 'high' BT milk ELISA at the 2017 follow up test. Farms that had calves born outside of calving areas were 2-4 times more likely to have successful JD control. With regards to weaned and bred heifer management, farms that had lower heifer cleanliness were more likely to have failure of JD control. The results indicate that it is not enough to say cows should be calved in designated calving pens. The management of these pens is critical in minimizing exposure of newborn calves to infectious material thereby reducing risk of JD transmission.

324

Invited Speaker for Diagnostics and Detection

Prof Vivek Kapur

Diagnostics and Detection - Invited Speaker, Main Auditorium, Printworks, June 14, 2022, 8:30 AM - 9:15 AM

Five learning points:

1. The control of Mycobacterium avium subsp. paratuberculosis (MAP) is achieved primarily through better management to reduce opportunities for introduction or within-herd transmission, regular testing to identify infected animals, and removal of infected individuals

2. In low-and-middle-income countries where the disease is endemic and dairy intensification is driving increase in prevalence, the testing and removal of MAP infected animals is economically and socially unfeasible

3. While commercial vaccines exist, vaccination for prevention (or treatment) of MAP infection in cattle is not widely practiced, in part because of the potential for interference with tuberculin-based testing for bovine tuberculosis

4. Newer generations of molecularly defined tests have been developed that show considerable promise in replacing tuberculin-based tests for TB diagnosis

5. Learning from the experience in the TB field, the next generation of successful MAP vaccines should consider companion diagnostics that enable differentiation of infected from vaccinated animals

248

Comparison of herd-level MAP prevalence based on F57 and ISMAP02 target genes

Mrs Larissa Martins[1], Dr. Karin Orsel[1], Dr. Jeroen De Buck[1], Dr. Herman Barkema[1]
[1]University of Calgary, Calgary, Canada

Diagnostics and Detection - AM Session (1), Main Auditorium, Printworks, June 14, 2022, 9:15 AM - 10:30 AM

Detection of Mycobacterium avium subsp. paratuberculosis (MAP) in dairy herds using environmental samples is a cost-effective surveillance method with adequate sensitivity (Se) and specificity (Sp). Although the F57 gene is MAP-specific, only one copy of this gene is present in the MAP genome, which could result in a lower Se of PCR. Improvements on the Se of MAP detection based on environmental samples might come from changing the MAP target gene. The aim of the study was to compare test accuracy of the ISMAP02 and F57 gene to detect MAP-positive environmental samples. Environmental samples were collected on 22 Alberta, Canada, dairy farms. DNA was extracted using the MagMax Total Nucleic Acid Isolation Kit (Thermo Fisher Scientific) followed by qPCR targeting F57 and ISMAP02 genes. Internal controls were developed to exclude potential false-negative results due to qPCR inhibition. On each farm, 3 samples were collected from the lactating cow and manure storage areas. Of the 132 environmental samples collected, 13.6% were positive for F57, whereas a high proportion of the samples were MAP-positive using ISMAP02 (84.1%). In lactating cow areas, only 15.5% of the samples were F57-positive, whereas more than 80% were ISMAP02-positive. The results were similar for manure storage areas with 12.1% and 86.4% of samples F57 and ISMAP02-positive, respectively. The high discrepancy between ISMAP02 and F57 genes will require analysis to exclude potential false-positive results yielded by ISMAP02 gene. Samples known to be ISMAP02-positive and negative will be provided by USDA for panel tests. In addition, different qPCR kits targeting ISMAP02 gene will be used and compared to F57 results. Additional MAP-specific genes, such as hspX and mbtA-MAP217, will be included in the analysis to optimize prevalence estimations based on direct PCR of environmental samples. The results of these further explorations will also be presented at the colloquium.

211

Evaluation of a droplet digital PCR (ddPCR) assay for detection and quantification of Mycobacterium avium subsp. paratuberculosis (MAP) DNA in whole-blood and fecal samples from MAP-infected Holstein cattle

Gerard Badia-Bringué[1], Maria Canive[1], Dr Rosa Casais[2], Cristina Blanco-Vázquez[2], Dr Javier Amado[3], Dr Natalia Iglesias[2], Dr Ramon A. Juste[1], Dr Marta Alonso[1]

[1]*Department of Animal Health, NEIKER- Basque Institute for Agricultural Research and Development, Basque Research and Technology Alliance (BRTA), Derio, Spain,* [2]*Center for Animal Biotechnology, Servicio Regional de Investigación y Desarrollo Agroalimentario (SERIDA), Deva, Spain,* [3]*Department of Microbiology, Laboratorio Regional de Sanidad Animal del Principado de Asturias (LSAPA), Gijón, Spain*

Diagnostics and Detection - AM Session (1), Main Auditorium, Printworks, June 14, 2022, 9:15 AM - 10:30 AM

Diagnostic tools for the direct detection of Mycobacterium avium subsp. paratuberculosis (MAP) are real-time PCR and bacteriological culture, the last one being considered the gold standard. However, both present several limitations to detect subclinical MAP-infected cattle with low bacterial load in feces and gut tissues. Droplet digital polymerase chain reaction (ddPCR) is a third-generation PCR method that shows high sensitivity for the detection and quantification of small DNA copy numbers. The objective of this study was to design a ddPCR assay to detect and quantify a fragment of the F57 MAP-specific sequence. DNA was isolated from whole-blood and fecal samples from ELISA and fecal PCR negative cows (N= 65), and from infected cows with PTB-associated focal (N=32), multifocal (N=21), and diffuse lesions (N=17) in gut tissues. After ddPCR, the fecal samples from the cows with multifocal (335.4) and diffuse (2,092,666.4) lesions showed higher mean copies/µl than the cows with focal lesions (114.1) or control cows (121.6) ($P \leq 0.05$). Significant differences in mean copies/µl were observed in the blood samples from cows with focal (49.7) and diffuse lesions (12.4) analyzed with ddPCR. When compared with a commercial real-time PCR, the fecal ddPCR was able to detect MAP DNA in 100 % of the positive PCR-positive samples and in 92.3 % of the PCR-negative fecal samples. When compared with other diagnostic tools, fecal ddPCR positively correlated with the results of a commercial ELISA for the specific detection of MAP antibodies, and with fecal and tissue PCR and bacteriological culture results. In contrast, blood ddPCR results negatively correlated with fecal PCR and fecal and tissue bacteriological culture results. In summary, our results demonstrate that the blood ddPCR could be used for MAP detection in cows with subclinical paratuberculosis with low amounts of feces and tissues.

316

Evaluation of the Actiphage® Rapid JD test as a pre-breeding heifer screening tool in endemically infected herds

Mr Timothy Geraghty[1], Mr George Caldow[1], Mr Graeme Richardson[2]
[1]SRUC, Aberdeen, United Kingdom, [2]Thrums Veterinary Group, Kirriemuir, United Kingdom

Diagnostics and Detection - AM Session (1), Main Auditorium, Printworks, June 14, 2022, 9:15 AM - 10:30 AM

Aim: To determine the usefulness of the Actiphage® Rapid JD test for screening of pre-breeding heifers in endemically infected herds as part of a control programme.

Methods: Nine commercial beef herds and one dairy (mean breeding cows 152; range 44-401) that had endemic Johne's disease infection and were actively engaged in Johne's control participated. Adult animal (>24 months old) Johne's antibody ELISA positive prevalence for the two years prior to enrolment ranged from 0.65%-8.7% (mean 3.8%). In April 2019, pre-breeding heifers were tested using a commercial Johne's disease ELISA (ID Screen® Paratuberculosis Indirect; ID Vet) and by the Actiphage® Rapid JD Test (PBD Biotech Ltd), with a view to excluding test positives (by either test) from breeding.

Results: 218 homebred heifers and 22 purchased heifers, age range 10-29 months, were sampled (7-48 per herd). All 240 tested negative by the ELISA test. In contrast, 103 heifers (43%) tested positive by the Actiphage® test. Test positive rate per herd ranged from 17-78%. There was a slight positive correlation between mean adult animal ELISA positive prevalence at the start of the study and the homebred heifer Actiphage® test positive prevalence (R2=0.22). 204 animals had at least 1 follow-up ELISA test in the following 2 years, of which 8 tested positive (3.9%). At the time of abstract submission there was no difference in the occurrence of subsequent ELISA positive test results between animals that had initially tested negative on Actiphage® (5/118; 4.2%), and those that had initially tested positive (3/81; 3.7%).

Conclusion: The Actiphage® Rapid JD Test was not considered a useful screening as the high positive rate made it impractical to exclude test positives from breeding. Further, the Actiphage® result did not correlate with occurrence of testing positive by antibody ELISA in the 2.5 years after the initial test.

256

Development of LAMP coupled LFD Kit for Rapid and Specific Detection of Mycobacterium avium subspecies paratuberculosis in Field

Dr Jagdip Singh Sohal[1], Dr Manju Singh[1], Dr Mukta Jain[1], Professor Shoor Vir Singh[2], Mr Amit Kumar Singh[1], Dr Saurabh Gupta[2], Dr Janardhan Reddy[3], Dr Rathnagiri Polavarapu[3], Dr Amit Kumar[4], Dr Deepansh Sharma[1], Professor G K Aseri[1]

[1]Amity Center For Mycobacterial Disease Research, Amity University Rajasthan, Jaipur, India, [2]GLA University, Mathura, India, [3]Genomix Molecular Diagnostics Pvt. Ltd., Hyderabad, India, [4]Sardar Vallabh Bhai Patel University of Agriculture & Technology, Meerut, India

Diagnostics and Detection - AM Session (1), Main Auditorium, Printworks, June 14, 2022, 9:15 AM - 10:30 AM

In various veterinary diseases Test and Cull or Test and Segregate policy is implemented to control the spread of disease. Thus, demonstrating the presence of pathogen in animals makes decision making easy. PCR based diagnosis is most definitive tool in infectious diseases like paratuberculosis. Loop-mediated isothermal amplification (LAMP) has far better detection limit compared to traditional PCR and qPCR. This study aimed to design and develop a LAMP coupled lateral flow device (pLAMP coupled LFD) to detect Mycobacterium avium subspecies paratuberculosis (MAP) in fecal samples for the onsite diagnosis. LAMP assay was first optimized on MAP genomic DNA and spiked fecal samples using six primers (two loop primers) targeting IS900 element. Detection limit of the optimized LAMP was 10 CFU/g of the feces and 10 fg MAP genomic DNA. For LFD based detection forward loop primer with biotin and backward loop primer with fluorescein isothiocyanate (FITC) were designed. LFD device contained a strip having distinct detection line (anti-FITC IgG) and control line. Development of purple reaction at both detection line and control line is indicative of positive reaction. Optimized LAMP assay was then used for onsite diagnosis in LAMP coupled LFD format. Onsite testing of 221 fecal samples from endemic herds (cattle, goat, sheep and buffalo) was done. Of these 221 animals; 65 (29.1%) were positive in LAMP coupled LFD. For confirmation agarose gel was also run the lab and identical findings were obtained. Compared to LAMP detection limit of traditional PCR and qPCR was lower, 22 (9.9%) and 52 (23.5%), respectively. There was absolute agreement in the performance of LAMP coupled LDF in third party validation and assay was user friendly. Findings of this study reveals the applicability of LAMP coupled LFD in resource poor areas.

213

Evaluation of the bovine ATP-binding cassette subfamily A member 13 (ABCA13) as a potential biomarker for sensitive detection of animals with focal pathological forms of subclinical paratuberculosis.

Dr. Rosa Casais[1], Miss Cristina Blanco-Vázquez[1], Miss Natalia Iglesias[1], Dr. Patricia Vázquez[2], Dr. R.A. Juste[2], Dr. J.M. Garrido[2], Dr. Ana Balseiro[3], Miss María Canive[2], Mister Javier Amado[4], Mister M.A. Queipo[5], Miss Tania Iglesias[6], Dr. Marta Alonso-Hearn[2]

[1]Centro de Biotecnología Animal, Servicio Regional de Investigación y Desarrollo Agroalimentario (SERIDA), 33394 Deva, Asturias, Spain., Gijón, Spain, [2]Department of Animal Health, NEIKER-Basque Institute for Agricultural Research and Development. Basque Research and Technology Alliance (BRTA). Parque Científico y Tecnológico de Bizkaia, P812, E-48160 Derio, Spain., Derio, Spain, [3]Departamento de Sanidad Animal, Facultad de Veterinaria, Universidad de León, 24071 León, Spain; Instituto de Ganadería de Montaña (IGM, CSIC-ULE), Finca Marzanas, Grulleros, 24346 León, Spain., León, Spain, [4]Laboratorio Regional de Sanidad Animal del Principado de Asturias, 33201 Gijón, Asturias, Spain. , Gijón, Spain, [5]Servicio de Sanidad y Producción Animal del Principado de Asturias, 33007 Oviedo, Asturias, Spain., Gijón, Spain, [6]Unidad de Consultoría Estadística, Servicios científico-técnicos, Universidad de Oviedo, Campus de Gijón, Asturias, Spain. , Oviedo, Spain

Diagnostics and Detection - AM Session (1), Main Auditorium, Printworks, June 14, 2022, 9:15 AM - 10:30 AM

Paratuberculosis (PTB) control continues to be a great challenge today, due, among other factors, to the low sensitivity of the current diagnostic techniques for detection of subclinical cases. RNA-Seq analysis showed that the ABCA13 mRNA was overexpressed (log2 fold change 3.74) in whole blood of animals with focal histological lesions, the most frequent histological form of subclinical PTB. In the present study, the potential as a sensitive diagnostic tool of an ELISA based on ABCA13 detection was evaluated in serum samples of 704 Holstein Friesian cows (566 naturally infected animals and 138 negative control animals from PTB-free farms). The infected animals were classified into three groups according to the type of histological lesions present in their intestine and associated lymph nodes: focal (n=444), multifocal (n=59) and diffuse (n=60). The results indicate that the ELISA based on the detection of ABCA13 showed good discriminatory power between animals with focal lesions and control animals with an AUC value of 0.869 ($p<0.001$), a sensitivity of the 82.99% and a specificity of 80.43%. However, its discriminatory power between animals with multifocal or diffuse lesions and the control animals was poor (AUC <0.7, $p<0.001$). The ability of the ABCA13-based ELISA to detect animals with focal lesions was compared with that of other PTB diagnostic methods showing higher sensitivity (82.99%) than the IDEXX ELISA (3.36%), the fecal bacteriological isolation (4.65%) and the real-time PCR (6.15%). The combined use of the ABCA13-based ELISA and the IDEXX ELISA, which has a better diagnostic performance for the detection of animals with diffuse lesions, increased global sensitivity for the detection of animals with any type of PTB-associated lesion from 77.56% to 83.03%. These results indicate that the ABCA13 ELISA greatly improves the identification of subclinical animals with focal lesions that are undetectable using conventional diagnostic methods.

197

A Novel 2-in-1 Culture System for the Isolation of Mycobacterium avium subsp. paratuberculosis

Miss Hannah Dane[1], Dr Linda Stewart[1], Professor Irene Grant[1]
[1]*Institute for Global Food Security, School of Biological Sciences, Queen's University Belfast, Belfast, United Kingdom*

Diagnostics and Detection - AM Session (2), Main Auditorium, Printworks, June 14, 2022, 11:00 AM - 12:45 PM

Culture remains the 'gold standard' detection method for viable Mycobacterium avium subsp. paratuberculosis (MAP). While liquid culture methods speed up detection time, culture on a solid medium is needed to obtain MAP colonies. The aim of this study was to develop a '2-in-1' culture system with liquid and solid media in the same flask, to potentially increase MAP colony numbers and colony formation rates compared to solid media alone. Tissue culture flasks (25 ml) were prepared with 6 ml agar (7H10, 7H11 or HEYM) and 2 ml broth (7H9 or Pozzato). Once MAP inoculum was added to the broth, flasks were incubated in a vertical position for 1-4 weeks, then incubated in a horizontal position for 24 h to inoculate the agar surface, before being returned to a vertical position for incubation until colonies appeared. Different liquid-solid media combinations and periods of incubation in broth were investigated. The optimum medium combination and incubation period in broth varied depending on the concentration of MAP added. When using a 10^1-10^2 cfu MAP inoculum, colonies formed more rapidly in flasks containing 7H9 broth with HEYM, and an incubation period in broth of 3-4 weeks resulted in significantly higher growth scores ($p \le 0.01$). However, a 10^4-10^5 cfu MAP inoculum became visible earlier in flasks containing Pozzato broth with 7H10 medium, and longer incubation periods merely extended the time required to observe colonies. To date, although the 2-in-1 system has not improved MAP colony formation rates compared to culture on agar alone, final colony counts were increased, and solid media didn't dry out. Experiments were completed with laboratory-grown MAP strains, so further work is required to optimise this novel 2-in-1 culture system and to determine whether colony formation rates upon primary isolation of MAP can be improved compared to culture on agar only.

210

More Insights about the Efficacy of Copper Ion Treatment on Mycobacterium avium subsp. paratuberculosis (MAP). A Clue for the Observed Tolerance

Mr Carlos Tejeda[1], Mss Marcela Villegas[1], Dr Pamela Steuer[1], Dr Miguel Salgado[1]
[1]*Instituto de Medicina Preventiva Veterinaria, Facultad De Ciencias Veterinarias, Universidad Austral De Chile, Valdivia, Chile*

Diagnostics and Detection - AM Session (2), Main Auditorium, Printworks, June 14, 2022, 11:00 AM - 12:45 PM

Introduction: Scientific evidence for the antimicrobial effect of copper-based treatments on microorganisms is extensive. However, less information is available for those bacteria characterized as more resistant, such as members of the Mycobacterium genus. Using Mycobacterium avium subsp. paratuberculosis (MAP), a highly resistant microorganism, as a pathogen model, copper ion treatment has shown a significant bactericidal effect both in buffer and milk, although some survival was also observed. With the aim of deepening our understanding of the efficacy of this novel treatment.

Material and Methods: A MAP spiked buffer matrix was treated using copper ions. We evaluated efficacy again on MAP cells but this time in conjunction with the physicochemical properties of the PBS buffer in which MAP cells had been suspended, since these could modulate MAP inhibition and / or its tolerance during the application of this novel copper treatment

Results: Despite the efficacy of copper ions in significantly reducing the MAP load in PBS buffer, some MAP cells were able to survive and grow in culture media. The copper concentration generated by the copper ion treatment device increased significantly with increasing exposure times. MAP bacterial load measured by qPCR decreased significantly when treated with copper ions as the exposure times increased, but this decrease was smaller in MGIT culture. An increase in pH, decreased oxygen consumption and an increase in conductivity was reported after the application of this copper-based treatment.

Conclusion: Even with higher concentrations of copper, the efficacy of MAP control was not complete. The concentration of copper must be a key element in achieving the high level of efficiency that we expected from this MAP control treatment.

299

Serum metabolome changes in bovine paratuberculosis by NMR spectroscopy and blood indices

Dr Nicola Pozzato[1], Dr Andrea Massaro[2], Dr Alessandra Tata[2], Dr Ivana Pallante[1], Massimo Bottazzari[3], Paola Fiorini[3], Dr Mauro Dal Prà[3], Dr Annalisa Stefani[4], Valentina Bertazzo[4], Prof Jeroen De Buck[5], Dr Roberto Piro[2]
[1]Istituto Zooprofilattico Sperimentale Delle Venezie, Laboratorio di medicina forense veterinaria, Vicenza, Italy, [2]Istituto Zooprofilattico Sperimentale Delle Venezie, Laboratorio di chimica sperimentale, Vicenza, Italy, [3]Istituto Zooprofilattico Sperimentale Delle Venezie, Laboratorio di Diagnostica Clinica e Sierologia di Piano, Verona, Italy, [4]Istituto Zooprofilattico Sperimentale delle Venezie, Laboratorio di Medicina di laboratorio, Legnaro, Italy, [5]Faculty of Veterinary Medicine, University of Calgary, Calgary, Canada

Diagnostics and Detection - AM Session (2), Main Auditorium, Printworks, June 14, 2022, 11:00 AM - 12:45 PM

Early detection of Mycobacterium avium subsp. paratuberculosis (MAP) is of paramount importance for farmers and veterinarians. Spectrometric and spectroscopic methods recently revealed metabolic changes in natural and experimental MAP-infected cattle, respectively. In this work, we aimed to investigate the changes in metabolic markers associated with natural MAP infection in infected and infectious bovines using quantitative nuclear magnetic resonance (NMR) spectroscopy and complete blood chemistry panel.
The prospective study included 23 infectious, 10 infected non-infectious/seronegatives and 26 negative animals. The blood indices and the data of the NMR were autoscaled and then statistically analyzed, separately, using R Studio software 3.6.2. Considering the low discrimination rates achieved, a data fusion approach was attempted. In details, each dataset was autoscaled and then concatenated by low-level data fusion resulting in a unique global fingerprint. Afterwards, the merged dataset was analyzed by the least absolute shrinkage and selection operator (LASSO). A mathematical weight for each statistically informative feature was calculated by the LASSO depending on the capability of a metabolite in characterizing a certain class. Based on the importance table of features, retrieved from LASSO, we observed that direct bilirubin was absent in negative cattle, but started to increase in infected cows and achieved the highest importance in infectious animals. On the other hand, lysin, cholesterol, malonate and 3-hydroxy-isovaleric acid received a positive weight in negative cows while not being informative for infectious and infected cows. A repeated k-fold cross-validation was performed for the model evaluation, achieving an overall accuracy of 91.5% with high values of sensitivity and specificity for each class. In conclusion, we were able to comprehensively explore the metabolic content, provided by the combination of the two analytical approaches, and achieve a good discrimination of the three groups that could be applied for early detection of MAP infected animals within infected herds.

218

The fate of heifers with 'low-positive' milk ELISA results

Dr. Maarten F. Weber[1,2], Dr. Marian Aalberts[1], Dr. Thomas Dijkstra[1], Prof. Dr. Ynte H. Schukken[1,2,3]
[1]Royal GD, Deventer, the Netherlands, [2]Dept. of Population Health Sciences, Faculty of Veterinary Medicine, Utrecht University, the Netherlands, , the Netherlands, [3]Quantitative Veterinary Epidemiology, Dept. of Animal Sciences, Wageningen University, the Netherlands, , the Netherlands

Diagnostics and Detection - AM Session (2), Main Auditorium, Printworks, June 14, 2022, 11:00 AM - 12:45 PM

Historically, low-positive antibody responses against Mycobacterium avium subsp. paratuberculosis (Map) in clinically healthy cattle were considered a sign of progression to advanced stages of the disease. More recent studies show that the rate of disease progression is highly variable. However, data on the course of infection in cattle with low antibody responses are lacking. Therefore, we studied the fate of heifers with a low-positive milk ELISA result.

Data obtained between 2008 and 2019 from 718 dairy herds participating in the Dutch milk quality assurance programme were analysed. Milk samples of lactating cattle were tested annually or biannually by ELISA. Only high-positive results (S/P≥1.00) were reported as positive given the high concurrent probability of faecal shedding of Map; all other samples were reported as negative. Thus, cattle with low-positive (0.20<S/P<1.00) results and cattle with truly negative results (S/P≤0.20) were likely to be managed similarly. The effects of the S/P of the first milk ELISA result of 90,835 heifers on time to becoming high-positive and on time to culling were analysed with Cox regression models, taking into account the interval censored nature of the data.

In heifers with an S/P of their first milk ELISA classified as S/P≤0.20, 0.20<S/P≤0.40, 0.40<S/P≤0.60, 0.60<S/P≤0.80 or 0.80<S/P<1.00, the fitted probabilities of becoming high positive (S/P≥1.00) within two years after the first milk ELISA were 2%, 8%, 16%, 18% and 51% respectively, whereas the overall probabilities of culling within two years were 30%, 38%, 38%, 47% and 67%, respectively. Thus, heifers with 0.80<S/P<1.00 are at high risk of disease progression. However, the rate of disease progression appears to be rather low in a considerable proportion of heifers with an 0.20<S/P≤0.80. Therefore, the frequently given advice to cull low-positive heifers based on their ELISA response alone (without subsequent confirmation of faecal shedding) needs to be reconsidered.

251

Development of a rapid multiplex immunoassay for the early detection of Mycobacterium avium paratuberculosis in bovine samples.

Dr Natasha Gordon[1], Dr Jimena Tejerina[1], Dr Dave Whittaker[1], Dr Richard Janeczko[1]
[1]*Pictor, Parnell, New Zealand*

Diagnostics and Detection - AM Session (2), Main Auditorium, Printworks, June 14, 2022, 11:00 AM - 12:45 PM

Johne's disease, caused by Mycobacterium avium paratuberculosis (MAP) infection, affects many bovine calves worldwide, resulting in a significant financial loss for the farming industry. The commercially-available in vitro diagnostic (IVD) kits are conventional ELISA tests that rely on a single antigen for the detection of the immune response towards MAP in bovine samples. These kits, however have poor sensitivity. With the use of multiplexing technology we have developed an improved immunoassay (IA) to detect anti-MAP antibodies in bovine samples. This IA is quicker, cheaper, and has increased sensitivity due to the detection of multiple biomarkers towards the MAP pathogen.

The new assay utilises an indirect IA format, which involves sequentially: microarray based immobilisation of different MAP antigens on a 96-well microtiter plate; blocking the non-specific protein binding sites; incubation with bovine samples; detection of anti-MAP antibodies in bovine samples by incubating with anti-bovine IgG labelled with horseradish peroxidase (HRP); incubation with precipitating TMB substrate; and readout via a benchtop colorimetric reader. The short overall assay processing time of ~1.5h facilitates rapid diagnosis of Johne's disease. Under optimized assay conditions, the developed MIA showed an improved sensitivity in comparison to the commercial kit used as a predicate. Large-scale testing has been scheduled to determine the IA's bioanalytical performance, which has enormous potential for effective monitoring and management of Johne's disease that is vital for the dairy industry.

229

Early detection of viable MAP shedding by dairy calves possible using a rapid and sensitive phagomagnetic separation (PhMS)-qPCR method

Miss Kayleigh Meek[1], Dr Sam Strain[2], Professor Niamh O'Connell[1], Professor Irene Grant[1]
[1]Queens University Belfast, Belfast, Northern Ireland, [2]Animal Health and Welfare Northern Ireland, Armagh, Northern Ireland

Diagnostics and Detection - AM Session (2), Main Auditorium, Printworks, June 14, 2022, 11:00 AM - 12:45 PM

A study was carried out to evaluate the potential utility of a novel phagomagnetic separation (PhMS)–qPCR assay to detect Mycobacterium avium subsp. paratuberculosis (MAP) infection in dairy calves early. Individual faecal samples were collected from 201 calves aged between 3 and 6 months in a UK dairy herd that has a high incidence of Johne's Disease infection. The faecal samples were pooled into 39 groups of 5 or 6 samples prior to PhMS-qPCR testing. Twelve of these pools (30.8 %) tested positive for viable MAP. Samples from each of the 63 faecal samples used in these pools were then tested individually. Viable MAP were detected by PhMS-qPCR in 28 (44.4%) of the 63 individually tested faecal samples, with estimated MAP contamination levels ranging from 1.0–1275.9 MAP/0.2g faeces. Three months later, when the calves were 6-9 months old, faecal samples were collected again from all individual calves that had previously tested PhMS-qPCR positive (n=28), and from a random selection of the calves that had previously tested PhMS-qPCR negative (n=14). Viable MAP was detected in 100% of faecal samples from the previously PhMS-qPCR positive calves, with MAP contamination levels ranging from 2.3–60.5 MAP/0.2g faeces. However, unexpectedly, 8 (57%) of 14 previously PhMS-qPCR negative calves tested positive. All faecal samples were decontaminated with 0.75% hexadecylpyridinium chloride for 24 h and cultured on Herrold's egg yolk medium, however culture results are still pending. Our findings suggest that PhMS-qPCR is a rapid and sensitive method of detecting viable MAP in faeces of calves at an earlier stage than is currently possible via faecal culture. However, whether the viable MAP detected represent environmental 'pass-through' or true MAP infection in these calves remains an unanswered question. We plan to test the calves again at 12-15 months old before they would be due to enter the main herd.

200

Phage-based Detection of Viable Mycobacterium avium subsp. paratuberculosis in Milk and Faeces Superior to Faecal qPCR and Serum-ELISA for Identifying Shedding Animals in a Northern Ireland Dairy Herd.

Prof Irene Grant[1], Dr Antonio Foddai[1], Dr Elizabeth Earle[2], Mr Jason Rankin[2], Mr Gary Watson[3], Dr Barry McInerney[4], Dr Sam Strain[5]

[1]Institute for Global Food Security, Queen's University Belfast, Belfast, United Kingdom, [2]The Northern Ireland Agricultural Research and Development Council (AgriSearch), Hillsborough, United Kingdom, [3]Dale Farm Limited, Belfast, United Kingdom, [4]Veterinary Sciences Division, Agri-Food and Biosciences Institute, Belfast, United Kingdom, [5]Animal Health and Welfare Northern Ireland, Dungannon, United Kingdom

Diagnostics and Detection - AM Session (2), Main Auditorium, Printworks, June 14, 2022, 11:00 AM - 12:45 PM

Between May and July 2019, blood and faeces from dairy cows in several Northern Ireland herds were tested by blood-ELISA (Mycobacterium paratuberculosis Antibody Test Kit, IDEXX) and faecal qPCR (VetMAX MAP Real-Time PCR screening kit, Applied Biosystems). In one of the 10 dairy herds sampled, all 60 lactating cows tested negative by both diagnostic tests applied, whereas in other herds variable numbers of test positive animals were encountered. Milk samples hand-milked from animals in the 'test negative' dairy herd (collected in parallel with the blood and faeces samples) were tested by our original PMS-phage assay, PMS-Pozzato culture, and a novel phage-based one-day test for viable MAP. Faeces were tested by the new one-day phage-based assay only. Surprisingly, 50% (30/60) of faeces samples and 33.3% (20/60) of milk samples yielded a positive result for the presence of viable MAP with the new one-day phage-based assay. The two other milk tests also yielded positive results – 18 of 60 (30%) and 32 of 60 (53.3%) milk samples tested positive for viable MAP by the original PMS-phage assay and PMS-culture, respectively. The mean numbers (± SD) of viable MAP indicated by the new one-day phage-based test for faeces and milk were 379.6 ± 238.9 PFU/g and 26.2 ± 5.4 PFU/50 ml, respectively. There was 'moderate' agreement (Kappa 0.467, 95%CI: 0.256-0.678, p=0.0001) between the one-day phage assay results for faeces and milk, with only four milk samples testing MAP positive in the absence of a positive result for the corresponding faeces sample. These findings may suggest that the two phage-based methods and culture after peptide-mediated magnetic separation (PMS) have greater MAP detection sensitivity than both faecal qPCR and blood-ELISA. Further work is required to investigate these findings. Neither phage-based test is a validated test for viable MAP in milk or faeces.

183

Mycobacteriophage characterisation: An important consideration for the biocontrol of Mycobacterium avium sbsp. paratuberculosis.

Miss Laura O'Connell[1], Prof. Jim O'Mahony[1]
[1]Munster Technological University, Bishopstown, Ireland

Diagnostics and detection - Poster Pitches (1), Main Auditorium, Printworks, June 14, 2022, 12:45 PM - 1:00 PM

Johne's disease is a chronic gastroenteritis condition that affects ruminants and often results in animal death and major economic losses. As a means to reduce the global impact of Johne's disease, it is important to develop accurate diagnostics which can rapidly identify diseased animals. Several diagnostic assays have been designed which involve the use of mycobacteriophages to detect Mycobacterium avium sbsp. paratuberculosis (MAP). In an effort to contribute to the available collection of mycobacteriophages, with the additional aim of identifying those capable of infecting various MAP strains, a simple screening study of environmental samples was conducted. Optimal sample storage conditions were defined with a spiking study prior to sample screening. Samples were washed with mycobacteriophage buffer which was then filtered and enriched to propagate any isolated phage using previously described methods. Traditional plaque and spot assays were used to confirm the presence or absence of mycobacteriophage in the enriched samples. The study was successful in isolating a novel Fishburnevirus, Nix22, from leaf litter collected in Phoenix Park, Co. Dublin, which demonstrates mycobactericidal activity against Mycobacterium smegmatis mc2155. Additional bioinformatic analyses are underway to better characterise Nix22 within the context of publicly available mycobacteriophage sequences and molecular experiments are planned to explore the functionality of genes which may benefit a diagnostic system. The ultimate goal is to determine the ability of Nix22 to recognise and infect MAP. Similar to the other mycobacteriophage in the Munster Technological University collection, Nix22 has the potential to be included in highly specific and sensitive diagnostic assays and highly targeted phage therapies against Johne's disease, which may greatly improve current control strategies, as well as animal welfare and agricultural profit margins.

267

Study on prevalence of Mycobacterium avium subspecies paratuberculosis (MAP) infection in suspected diarrhoeic buffaloes and cattle reporting at clinics for therapeutic intervention

Dr. Shalini Sharma[1,3], Dr. Ajay Kumar Gautam[1], Dr. Shoor Vir Singh[2], Dr. Kundan Kumar Chaubey[2], **Dr. Sujata Hariharan[2]**

[1]Department of Veterinary Physiology and Biochemistry, Lala Lajpat Rai University of Veterinary and Animal Sciences, Hisar-125004, Haryana, India, Hisar, India, [2]Department of Biotechnology, Gla University, Mathura, Mathura, India, [3]Department of Veterinary Clinical Complex, Lala Lajpat Rai University of Veterinary and Animal Sciences, Hisar-125004, Haryana, India, Hisar, India

Diagnostics and detection - Poster Pitches (1), Main Auditorium, Printworks, June 14, 2022, 12:45 PM - 1:00 PM

Background: Murrah buffaloes, best breed for milk production are native of Haryana state. They contributes significantly to the farmer's income, livelihood and food (milk and meat) security, in the semi-tropical regions of North India. Johne's disease though endemic in the domestic livestock of the country, but reports are not available on the status of JD in diarrhoeic buffaloes.

Method: We estimated the status of JD in diarrhoeic buffaloes and cattle reporting at Veterinary Clinical Complex of LalaLajpatRai University of Veterinary and Animal Sciences, Hisar, Haryana, India, using conventional, serological and PCR assays.

Results: 141 buffaloes suffering from chronic diarrhoea were screened to estimate sero-prevalence of MAP and 50.0 % young and 53.52 % adult animals were positive. Of 14 cattle screened, none of the young and 66.6 % adult cows were positive. In buffaloes, 66.1 and 6.77 %, fecal samples were positive in microscopy and IS900 PCR, respectively. Sero-prevalence of JD was very high in diarrhoeic buffaloes and cattle from Haryana state of India. Diminished TLC and lymphocytes were observed in MAP positive buffaloes.

Conclusions: High sero-prevalence of JD in diarrhoeic buffaloes signals towards the clinical stage of disease. Diminished leucocyte counts might reflect chronic stages of the disease and a reduced capability to handle the antigen.

300

Detection of MAP and M. avium co-infection in North American Bison using Actiphage

Mr Wade Handly Hartill[1], Ms Catherine Evans[1,3], Mr Albert Hunyh[1], Dr Tania Perehinec[1,3], Dr Benjamin Swift[2], **Prof Catherine Rees[1]**

[1]University of Nottingham, School of Biosciences, Sutton Bonington Campus, United Kingdom, [2]Royal Veterinary College, Hawkshead Campus, North Mymms, Hatfield, United Kingdom, [3]PBD Biotech Ltd, Thurston, United Kingdom

Diagnostics and detection - Poster Pitches (1), Main Auditorium, Printworks, June 14, 2022, 12:45 PM - 1:00 PM

Actiphage® can detect low numbers of Mycobacteria in blood samples. Since the assay directly detects the presence of the organism rather than an immune response, the method can be applied to any species of animal, and has been used to detect MAP in the blood of cattle and farmed deer. In this study, Actiphage was used to test samples from seven North American Bison that were suspected to have a mycobacterial infection. As the nature of the infection was unknown, different PCR assays were applied to identify the type of Mycobacterium causing the infection.

To perform the assay peripheral blood mononuclear cells (PBMCs) were purified from heparinised bison blood (7 animals) using Ficoll density gradients (four replica 2 ml blood samples from each animal). The purified PBMC's were lysed to release any intracellular mycobacteria and then tested using the Actiphage assay to recover mycobacterial DNA.

A PCR assay targeting the IS1311 element gave a positive PCR result (amplification of a 600 bp PCR product) indicating that M. avium Complex (MAC) organisms were detected in the blood samples. To identify which sub-type of the (MAC) detected, restriction enzyme analysis (REA) of PCR products was performed using the enzymes HinfI and MseI. In six animals, the results indicated that M. paratuberculosis was detected. Two of these animals also had evidence of co-infection with M. avium. In the remaining animal only DNA from M. avium was detected. Identical results were gained when the analysis was repeated with two replica samples. This is the first report where Actiphage has been used to detect MAP in North American Bison and the first time that co-infection of animals has been detected using this method. It also demonstrates the potential for direct sub-typing of organisms from blood without the need for primary culture.

310

Optimization of Actiphage for detection of MAP in different hosts, including humans

Dr Benjamin Swift[2], Mr Wade Handley-Hartill[1], Mr Anton Kubala[1,3], Ms Catherine Evans[1,3], **Prof Catherine Rees[1]**
[1]University of Nottingham, Loughborough, United Kingdom, [2]The Royal Veterinary College, Hawkshead Lane Campus, North Mymms , United Kingdom, [3]PBD Biotech Ltd, Thurston, United Kingdom

Diagnostics and detection - Poster Pitches (1), Main Auditorium, Printworks, June 14, 2022, 12:45 PM - 1:00 PM

Actiphage uses bacteriophage as the lysing agent to extract DNA from mycobacterial cells. The robust nature of the mycobacteria means that physical or chemical lysis methods are inefficient. In contrast, the phage lysis has evolved to be highly efficient and allows Actiphage to detect very low (<10) numbers of cells. Experiments were performed to optimise Actiphage for blood from host species other than cattle.
Methods used for purifying PBMCs including Ficoll density gradients, differential lysis (ACK buffer) and differential sedimentation (Hetasep). Different PCR formats (end-point and real-time) were compared.
Blood samples from domestic goats were tested using the standard Actiphage method but produced poor recovery of PBMCs which was overcome using the ACK lysis method. PCR sensitivity using primers P90 and P91 was poor, especial using a real-time PCR format. Comparison with MAP genome sequences found a sequence mismatch at the 3' end of the P90 primer in many non-cattle strains. A redesigned P90 was used to screen caprine blood samples of known ELISA status, and 60% of ELISA-positive animals gave a positive Actiphage test result, confirming the infection status of these animals. In a set of vaccinated animals, 1 was identified Actiphage positive, indicating that Actiphage can be used as a DIVA test for this species.
For human blood, Hetasep was found to be the best method to purify PBMCs. Samples were obtained from patients referred to a TB clinic with atypical symptoms. After Samples screening for the presence of MTBC DNA, any negative samples (=5) were then screened for the presence of MAC organisms using IS1311 PCR-REA. This analysis revealed that 1 patient had detectable levels of an S-strain of MAP in their blood.
These results demonstrate that some optimisation is required before Actiphage can be successfully applied to blood samples from different host species.

321

Invited Speaker for Host Response and Immunology

Dr Kumudika De Silva

Host Response and Immunology - Invited Speaker, Main Auditorium, Printworks, June 15, 2022, 8:30 AM - 9:15 AM

Five learning points

1. Exposure to MAP can result in a spectrum of outcomes ranging from severe clinical disease to resilience to infection

2. Early immune responses can be used to identify animals that have the potential to withstand disease progression

3. Classification of disease outcome is important when assessing biomarkers for protective immunity

4. A combination of cellular and molecular biomarkers is required to define resilience to paratuberculosis

5. Immunological profiles of resilience could be used for diagnostics as well as for improving vaccine design

234

An in vivo vaccine screening model in neonatal calves to investigate parenteral vaccine-induced mucosal immune responses to an enteric MAP infection

Dr Antonio Facciuolo[1], Dr. Amy Lee[2], Neil Rawlyk[1], Dr. Scott Napper[1,3], Dr. Andrew Potter[1], Dr. Volker Gerdts[1], Dr. Hugh Townsend[1], Dr. R.E.W. Hancock[4], Dr. Philip Griebel[1,5]

[1]*Vaccine & Infectious Disease Organization (VIDO), University of Saskatchewan, Saskatoon, Canada,* [2]*Department of Molecular Biology & Biochemistry, Simon Fraser University, Burnaby, Canada,* [3]*Department of Biochemistry, Microbiology and Immunology, University of Saskatchewan, Saskatoon, Canada,* [4]*Centre for Microbial Disease and Immunity Research, Department of Microbiology & Immunology, University of British Columbia, Vancouver, Canada,* [5]*School of Public Health, University of Saskatchewan, Saskatoon, Canada*

Host Response and Immunology - AM Session (1), Main Auditorium, Printworks, June 15, 2022, 9:15 AM - 10:30 AM

A major limitation in the MAP vaccine pipeline is an infection model that provides a consistent and quantitative measure of disease protection in the natural host. Intestinal segments were surgically isolated in calves to facilitate targeted delivery of a defined dose of MAP to Peyer's patches (PP). We confirmed this model provided reproducible infection by quantifying viable MAP in PP tissue and intestinal contents. We then tested the model by parenterally injecting calves with a licensed MAP bacterin (Silirum™) to determine if vaccine-induced immune responses were capable of controlling an enteric MAP infection. In two independent studies, one-month old calves were injected subcutaneously with either vaccine carrier (Controls; Trial 1: n=5; Trial 2: n=9) or Silirum™ (Trial 1: n=4; Trial 2: n=8). Two months post-vaccination, two intestinal segments were surgically isolated in each calf - one containing a discrete PP (DPP) and the other a continuous PP (CPP) - and each segment was injected with 10e9 MAP CFU. At 28 days post-infection, we quantified viable MAP in PP tissues and luminal contents from each intestinal segment and isolated mucosal leukocytes for immunological analyses. In both vaccine trials, MAP burden was significantly ($p < 0.05$) reduced relative to controls in CPP tissue and luminal contents, but not in DPP tissue and associated luminal contents. Flow cytometric analysis of mucosal leukocytes, MAP-specific cytokine responses, and RNA-Seq analysis revealed novel cytokine and cellular mucosal immune responses correlating with a reduced MAP burden in CPP but not DPP. Thus, targeted delivery of MAP to intestinal segments provides a quantitative and reproducible measure of infection and vaccine efficacy. Further, this model affords the unique opportunity to analyze vaccine-induced mucosal immune responses at the site of infection, and offers the MAP community a relatively rapid and quantitative approach for screening vaccine candidates in the natural host.

276

A genome-wide association study for tolerance to bovine paratuberculosis identifies genetic regions with candidate genes involved in DNA packaging, DNA damage response and innate immunity

Dr. Marta Alonso[1], Ms Maria Canive[1], Mr Gerard Badia-Bringué[1], Dr Patricia Vázquez[1], Dr Oscar González-Recio[2,3], Dr. Almudena Fernández[2], Dr Joseba Garrido[1], Dr. Ramon Juste[1]

[1]*NEIKER - Basque Research and Technology Alliance (BRTA), Derio, Spain,* [2]*Departamento de Mejora Genética Animal, Instituto Nacional de Investigación y Tecnología Agraria y Alimentaria, CSIC, Madrid, Spain,* [3]*Departamento de Producción Agraria, Escuela Técnica Superior de Ingeniería Agronómica, Alimentaria y de Biosistemas, Universidad Politécnica de Madrid, Ciudad Universitaria, Madrid, Spain*

Host Response and Immunology - AM Session (1), Main Auditorium, Printworks, June 15, 2022, 9:15 AM - 10:30 AM

Bovine paratuberculosis (PTB) has an important genetic component, with some animals being susceptible, resistant, or tolerant. Tolerant individuals are Mycobacterium avium subsp. paratuberculosis (MAP) infected Holstein cows with controlled pathology. In this study, the associations between host genetic and tolerance to MAP were explored using the genotypes from 277 Spanish Holstein cows with two distinct phenotypes: cases) positive PCR and culture result and with no lesions in gut tissues (N= 24) and controls) negative PCR and culture result and with PTB-associated lesions (N=253). DNA from peripheral blood of these animals was genotyped with the Bovine EuroG MD Bead Chip, and the corresponding genotypes were imputed to whole-genome sequencing (WGS) data using the 1,000 Bull genomes reference population. A genome-wide association study was performed using the WGS data and the defined phenotypes in a case-control approach. A total of 142 single nucleotide polymorphisms (SNPs) were associated (FDR ≤ 0.05, P < 5 x 10-5) with tolerance (heritability= 0.55). The 40 SNPs with the lowest P-value (P < 5 x 10-7) defined 9 QTLs located on BTA4, BTA9, BTA16, BTA25, and BTA26. Some of the identified QTLs overlapped with QTLs previously associated with PTB, bovine tuberculosis, mastitis, somatic cell score, bovine diarrhea virus persistent infection, tick resistance, and length of productive life. Functional analysis on the 98 candidate genes that overlapped with the identified QTLs revealed a significant enrichment of the DNA packaging and DNA damage response (DDR) (ITNS7/TNP2/PRMI1/PRM2/PRM3). In addition, the toxoplasmosis (bta05145; TGFβ2/CHUK/CIITA/SOCS1) and TNF-signaling (bta04668; TRAF5/CREB5/CASP7/CHUK) pathways were enriched. Interestingly, the nuclear Factor NF-Kappa-B Inhibitor Kinase Alpha (CHUK or IkBKA), a key molecule in the negative feedback of NF-kappa-B canonical signaling, was enriched in both pathways. Taken together, our findings provide evidence of a pro-homeostatic role of DNA damage-induced innate immune responses.

250

Dynamics of Immune Responses Following Immunization with a Live Attenuated Vaccine against Johne's disease.

Prof Adel Talaat[1], Mostafa Hanafy[1]
[1]University Of Wisconsin-madison, Madison, US, [2]Pan Genome Systems, INC., Madison, US

Host Response and Immunology - AM Session (1), Main Auditorium, Printworks, June 15, 2022, 9:15 AM - 10:30 AM

Johne's disease (JD) is a contagious, chronic, and potentially fatal infection caused by Mycobacterium avium ss. paratuberculosis (M. ap). Currently, there is no approved vaccine against JD in the USA. Previously, we showed that a single dose live attenuated vaccine (LAV) developed after deleting the lipN gene (dubbed pgsN) provided significant protection against JD in cattle. To decipher the dynamics of the immune responses elicited by the pgsN vaccine, we analyzed key parameters of goat immune responses following immunization with the live attenuated pgsN vaccine compared to the inactivated Mycopar® vaccine. In general, the pgsN vaccine was safe for administration via both intranasal (IN) and subcutaneous (SC) inoculations while the inactivated vaccine triggered granuloma formation in animals that lasted until the end of the experiment at 6 months post vaccination (MPV). Interestingly, immunohistochemistry of the site of inoculation indicated that pgsN was able to recruit more T lymphocytes while the inactivated vaccine recruited more B lymphocytes, a further indication of how each vaccine is developing immunity against JD. However, cell-mediated immune responses (as indicated by IFN-? release assay and intradermal comparative skin test) were prominent in both pgsN and Mycopar groups only when vaccines were given by SC route (not the IN route), with early induction following immunization only in the pgsN vaccine group. Animals in the inactivated vaccine group significantly reacted to purified protein derivatives (PPD) from M. bovis while animals received the pgsN vaccine group remained unchanged, suggesting cross reactivity of Mycopar with bovine tuberculosis antigens. Overall, the results of this study provided better understanding of the mechanism of the immune responses elicited by live versus inactivated vaccines and provided an alternative immunization route (e.g. IN) that could be further exploited to develop protective immunity against Johne's disease.

233

Regionally distinct mucosal immune responses to MAP infection in discrete and continuous Peyer's patches of young calves at 2 and 12 months post-infection

Dr Antonio Facciuolo[1], Dr. Amy H. Lee[2], Dr. Patricia Gonzalez Cano[3], Dr. Scott Napper[1,6], Dr. Andrew Potter[1], Dr. Volker Gerdts[1], Dr. R.E.W. Hancock[4], Dr. Lucy M. Mutharia[5], Dr. Philip Griebel[1,7]

[1]Vaccine & Infectious Disease Organization (VIDO), University of Saskatchewan, Saskatoon, Canada, [2]Department of Molecular Biology and Biochemistry, Simon Fraser University, Burnaby, Canada, [3]Department of Pharmacobiology, Universidad de la Canada, Oaxaca, Mexico, [4]Centre for Microbial Diseases and Immunity Research, Department of Microbiology & Immunology, Vancouver, Canada, [5]Department of Molecular & Cellular Biology, University of Guelph, Guelph, Canada, [6]Department of Biochemistry, Microbiology and Immunology, University of Saskatchewan, Saskatoon, Canada, [7]School of Public Health, University of Saskatchewan, Saskatoon, Canada

Host Response and Immunology - AM Session (1), Main Auditorium, Printworks, June 15, 2022, 9:15 AM - 10:30 AM

The ruminant small intestine contains two functionally and structurally distinct Peyer's patches (PPs). There are 25-40 discrete PPs (DPP) distributed throughout the jejunum and a single continuous PP (CPP) occupies the terminal jejunum and ileum. Mucosal immune responses to MAP infection occurring in PP are poorly understood and exclusively focused on the CPP. Using surgically isolated intestinal segments we targeted a defined dose of MAP (1 x 10e9 CFU) to individual PPs and demonstrated equal levels of MAP infection in both DPP (n=3) and CPP (n=3) at two months post-infection (PI). MAP infection induced antigen-specific IgA B cell responses in DPP but not in CPP, revealing both a regional and functional dichotomy in the mucosal immune response to MAP. This led us to hypothesize that adaptive immune responses induced in DPP may control MAP infection. At 12 months PI, MAP persisted in all segments containing a CPP (n=4), but was detected in only two of 5 segments containing a DPP (n=5). RNA-seq analysis revealed MAP persistence in CPP coincided with minimal transcriptomic changes (4 differentially expressed [DE] genes) whereas control of infection in DPP coincided with extensive transcriptomic changes (1,707 DE genes). This comparative analysis of DPP and CPP reveals a marked dichotomy in MAP persistence while transcriptional profiling reveals novel host responses associated with the control of MAP infection. Subsequent analysis of cytokine gene expression in discrete and continuous PP tissue, and MAP-specific recall responses in mucosal leukocytes isolated from gut-associated lymphoid tissues (PP, lamina propria and mesenteric lymph node) revealed IL22 and IL27 as correlates of protection. These studies show that MAP infection induces regionally-distinct mucosal immune responses in the small intestine, and provides a new perspective on the early stages of the host-pathogen interaction in the natural host.

201

Characterising immune responses following vaccination of calves with live-attenuated strains of Mycobacterium avium subsp. paratuberculosis

Miss Razieh EshraghiSamani[1], Dr/ Professor Jeroen De Buck[1]
[1]Department of Production Animal Health, Faculty of Veterinary Medicine, University Of Calgary, Calgary,, Canada

Host Response and Immunology - AM Session (1), Main Auditorium, Printworks, June 15, 2022, 9:15 AM - 10:30 AM

Johne's disease vaccines partially reduce the fecal shedding and postpone the incidence of clinical symptoms. Our study aims to develop a vaccine that overcomes the shortcomings of previous vaccines. Essential genes for MAP survival in the host's body have been identified by Tn-seq. Two essential genes were chosen for further studies, which play a role in iron acquisition (IA) and fatty acid metabolism (FAM), respectively. Live-attenuated vaccine strains were created by special transduction. Twenty-three animals were randomized over four groups including noninfected control and wild-type (WT) strain. Calves were inoculated with a dose of 109 at two weeks of age. Blood samples were collected at 12-, 14- and 16-week post inoculation for ex-vivo MAP viability assay and to study relative changes in cytokine transcription in PBMCs stimulated with PPD. Both WT and IA upregulated Th1/proinflammatory and Th17 cytokines. However, significantly higher levels of INFG, IL-12 and IL-17 were detected in the IA group at 12-week post inoculation. IL-10 expression did not differ significantly between wild-type and mutant strains. For viability assay, WT infected macrophages from each group were incubated with stimulated PBMCs, non-stimulated PBMCs, and medium alone. The percentage of viable MAP recovered from macrophages were detected using F57 qPCR after treatment with DNA dye. IA and FAM had significantly lower percentages of viable MAP recovered from macrophages incubated with unstimulated macrophages. All groups recovered the same percentage of viable cells from macrophages incubated with stimulated PBMCs. Overall, the natural host have been used to gain a better understanding of the immune response to Map and mutant strains. IA showed promise as a vaccine strain by inducing proinflammatory cytokine expression and ability to kill MAP in ex-vivo viability assay.

184

Identification of genetic regions associated with a low Mycobacterium avium subsp. paratuberculosis load within infected monocyte-derived macrophages

María Canive[1], Gerard Badia-Bringué[1], Dr Marta Alonso-Hearn[1]
[1]Neiker, Insituto Vasco de Investigación y Desarrollo Agrario, Derio, Spain

Host Response and Immunology - AM Session (2), Main Auditorium, Printworks, June 15, 2022, 11:00 AM - 12:15 PM

Mycobacterium avium subsp. paratuberculosis (MAP) spends most of its life cycle within macrophages. Since macrophages´ s functions are controlled by a limited number of genes under a controlled environment, the probability of identifying animals with a superior immune response against MAP is much higher using ex vivo macrophages' s models. In the current study, we characterized MAP-infected monocyte-derived macrophages (MDMs) isolated from peripheral blood of 75 Holsteins cows and infected ex vivo with MAP. Bacterial load (logCFUs) was quantified at 2h and 7 days p.i. using a Bactec MGIT 960 instrument. In addition, the levels of Epiregulin (EREG), Complement C3 (C3), galectin-9 (LGALS9), and nitric oxide (NO) were measured in the supernatant of the infected cells at 2h p.i. DNA from peripheral blood samples of the animals included in the study were genotyped with the EuroG MD bead Chip (44,779 single nucleotide-polymorphisms, SNPs). Linear mixed models were used to calculate the heritability (h2) estimates for each phenotype. After performing a genome-wide association study, the only phenotype that showed SNPs with a significant association (PFDR ≤0.05) was the bacterial load within MDMs. A total of 6 QTLs and 60 candidate genes located on BTA2, BTA17, BTA18, and BTA21 were associated with the estimated logCFUs at 2h p.i. Overlap was seen in two QTLs associated with the logCFUs at 2h (h2= 0. 87) and 7 d. p.i. (h2= 0.83). All the SNPs had negative regression coefficients, and were, therefore, associated with a low bacterial load within MDMs. Pathway analysis with candidate genes overlapping the identified QTLs revealed a significant enrichment of the apoptotic and calcium signaling pathways at 2 h p.i. and 7 days p.i, respectively. The identified QTLs overlapped with QTLs associated with bovine respiratory disease, somatic cell score, gastrointestinal nematode burden, tick resistance, IgG levels, and length of productive life.

203

Understanding the Invasion Dynamics of Mycobacterium Avium subsp. Paratuberculosis (MAP)

Mrs. Grace Baruta[1], Dr. Hong Zhang[1], Mrs. Laurie Alston[1], Dr. Jeroen De Buck[1], Dr. Simon Hirota[1]
[1]*University Of Calgary, Calgary, Canada*

Host Response and Immunology - AM Session (2), Main Auditorium, Printworks, June 15, 2022, 11:00 AM - 12:15 PM

Background: Given the economic burden associated with Johne's Disease (JD), considerable efforts have sought to understand MAP invasion dynamics. However, these processes have not been completely characterized hindering our ability to generate anti-infective agents. While the current paradigm suggests MAP travels to the small intestine, where it gains entry through the epithelium, the exact cellular tropism/ mechanism(s) of entry are not well defined. Therefore, we have developed an ex vivo enteroid system to visualize invasion of MAP in distinct cells of the intestinal epithelium using a GFP-expressing MAP strain. We sought to test the hypothesis that MAP invasion occurs via M cells through receptor mediated transcytosis.

Aims: 1) To characterize murine enteroid system and visualize MAP invasion 2) Determine cellular tropism for MAP within the mouse intestinal epithelium and 3) to uncover mechanisms underlying MAP invasion

Methods: Murine enteroids were generated, and M cell differentiation induced by the addition of RANKL. Ileal monolayers were exposed to GFP-expressing MAP (K-10 pWES4). Confocal microscopy was performed, and barrier function measured via trans-epithelial electrical resistance (TEER). GFP-MAP was incubated with fibronectin or fibronectin with integrin-blocking peptides prior to exposure to 2D monolayers.

Results: We generated enteroid monolayers with functional M cells capable of transcytosis, as demonstrated via mature M cell markers and active bead uptake. MAP was detected mainly within M cells, more notably with the fibronectin pre-incubation. Integrin-blocking peptides decreased this M cell tropism. Alterations in TEER following MAP exposure suggest the existence of a novel mechanism by which MAP disrupts the barrier to invade the mucosa.

Conclusion: The newly optimized approach provides the experimental system that will enable us to better characterize M cell-MAP interactions with the hopes of identifying new therapeutic targets to prevent the spread of MAP and reduce the economic impact of JD.

223

Uncovering miRNA signatures of resilience to paratuberculosis infection and their functional relevance in disease progression

Dr Kathryn Wright[1,2], Dr Kumudika de Silva[1], Dr Auriol Purdie[1], Dr Karren Plain[1], Dr Stefan Oehlers[2,3,4]

[1]Sydney School of Veterinary Science, The University Of Sydney, Camden, Australia, [2]Tuberculosis Research program, The Centenary Institute, Camperdown, Australia, [3]Discipline of Infectious Diseases and Immunology, The University of Sydney, Camperdown, Australia , [4]Marie Bashir Institute, Camperdown , Australia

Host Response and Immunology - AM Session (2), Main Auditorium, Printworks, June 15, 2022, 11:00 AM - 12:15 PM

Early detection of infection using current methods is limited due to the lengthy silent subclinical phase of Johne's disease (paratuberculosis), where animals may be shedding viable bacteria intermittently. Biomarkers are a potential diagnostic tool for early detection of MAP infection. microRNA (miRNA), a non-coding subset of RNA, have been proposed as biomarkers as they are stable and regulate gene expression. miRNA are promising biomarkers for the diagnosis, and differentiation between paratuberculosis disease states. Identification of animals which do not progress to clinical infection despite encountering an infectious dose of MAP, termed resilient animals, may also be possible using miRNA.

To assess the efficacy of miRNA as biomarkers for paratuberculosis, plasma samples from MAP infected sheep (n=24) were analysed using small RNA sequencing. A total of 58 miRNA were found to be differentially expressed (fold change > ±1.5, FDR 0.1, $q < 0.05$) in MAP infected animals (culture positive post-mortem lesions), and 25 miRNA differentially regulated in resilient animals. The miRNA dataset was further analysed using bioinformatics alongside published datasets to provide full transcriptomic profiles of infected and resilient animals.

To further assess the biological relevance of these dysregulated miRNA and how they contribute to resilience and infection, we utilised a zebrafish mycobacteriosis model. Zebrafish present an ideal organism to investigate mycobacterial pathogenesis, with a functional innate immune system during early embryogenesis. miR's -126a and -206 were selected from the screen for their expression in infected or resilient sheep. Target genes were analysed to identify altered pathways and mechanisms of pathogenesis. Using Crispr-Cas9 and transgenic embryos, host-pathogen interactions were able to be visualised and quantified, uncovering miRNA-dependant alterations of host neutrophil and macrophage responses that impact disease outcome. Using both ruminant and zebrafish models has allowed for progression from large-scale transcriptomic data to functional analysis in a cost-effective and efficient pipeline.

208

Alternative splicing of pre-mRNA modulates the immune response in peripheral blood and gut tissues of Holstein cattle naturally infected with Mycobacterium avium subsp. paratuberculosis

Gerard Badia-Bringué[1], Maria Canive[1], Dr Jose Luis Lavín[2], Dr Rosa Casais[3], Cristina Blanco-Vázquez[3], Dr Marta Alonso-Hearn[1]

[1]Department of Animal Health, NEIKER- Basque Institute for Agricultural Research and Development, Basque Research and Technology Alliance (BRTA), Derio, Spain, [2]Department of Applied Mathematics, NEIKER- Basque Institute for Agricultural Research and Development, Basque Research and Technology Alliance (BRTA), Derio, Spain, [3]SERIDA, Servicio Regional de Investigación y Desarrollo Agroalimentario, Center of Animal Biotechnology, Deva, Spain

Host Response and Immunology - AM Session (2), Main Auditorium, Printworks, June 15, 2022, 11:00 AM - 12:15 PM

Alternative splicing (AS) is an important mechanism of gene expression regulation that can influence pre-mRNA stability or structure, function, and localization of corresponding proteins with important physiological consequences. However, the role of AS in regulating the host response to Mycobacterium avium subsp paratuberculosis (MAP) infection is still unclear. In the current study, differential AS was analyzed using RNA-Seq data from peripheral blood (PB) and ileocecal valve (ICV) samples from Holstein cattle without lesions (N=4) and with focal (N=5) or diffuse (N=5) PTB-associated lesions in gut tissues. Using the RNA-Seq mapped reads, a multivariate analysis of AS events was performed with rMATS 3.2. rMATS uses a likelihood-ratio test to confirm that the difference in the mean exon inclusion levels between two sample groups exceeds a given threshold (>5%). In the PB samples of the infected animals, several genes of the neutrophils degranulation pathway (CD53, CLEC12A, CPNE1, ITGA2B, NCF1, SGSH, SIRPA, SLC11A1, TARM1, TMC6) and clathrin-mediated endocytosis (CLTA, DNM1, DNM2, EPS15L1, FCHO1) showed significant AS perturbations upon MAP infection. In the ICV samples of the infected cows, proteins with RNA-recognition and coiled-coil domains showed differential AS events when compared with control cows. More specifically, in the ICV from animals with focal lesions, several genes of the innate immune response (C2, C4A, CD46, CFH, CYLD, IRF3, PYCARD, TMEM173, TRIM38) showed differential AS when compared with cows without lesions. Changes in the AS of several genes correlated with changes in gene expression. In the ICV of animals with diffuse lesions, for instance, several components of the immune response (CLEC7A, BOLA, PSBM10, IFI30, IRF5, ARID5A, IL7) showed deregulation both in AS and mRNA expression. AS impacted the expression of several genes (BOLA, MHCI-A and BOLA-NCI) associated with the antigen presentation pathway and with several human autoimmune diseases.

315

Identification of disease tolerant and susceptible cows to Mycobacterium avium subspecies paratuberculosis infection for genetic analyses of disease resistance

Dr Nathalie Bissonnette[1], Scientific Researcher, Veterinarian Dave Kelton[2], Scientific Researcher, Veterinarian Gilles Fecteau[3], Scientific Researcher Kapil Tahlan[4], Scientific Researcher Eveline Ibeagha-Awemu[1], Scientific Researcher Philip Griebel

[1]Sherbrooke Research and Development Center, Agriculture and Agri-Food Canada, Sherbrooke, Canada, [2]University of Guelph, Guelph, Canada, [3]Faculté de médecine vétérinaire, Université de Montréal, Saint-Hyacinthe, Canada, [4]Department of Biology, Memorial University of Newfoundland, St. John's, Canada, [5]School of Public Health, University of Saskatchewan, Saskatoon, Canada

Host Response and Immunology - AM Session (2), Main Auditorium, Printworks, June 15, 2022, 11:00 AM - 12:15 PM

Following infection by Mycobacterium avium subspecies paratuberculosis (Map), some cows progress to clinical Johne's disease (JD) while others acquire tolerance to the disease. Disease progression appears unpredictable in large part because of individual host variation in resistance. Host genetic variation contributes to resistance and retarded progression toward clinical stages, i.e. tolerance. To identify genetic markers associated to tolerance (i.e. cows with the ability to cope with the infection), phenotypes of fecal Map-excretion and Map-specific blood antibodies (Map-Ab) were collected from cows older than 2 yrs for a period of 3-5 yrs. All cows came from herds with endemic MAP infection. Blood and fecal samples were collected 2 times/year and tested for Map-Ab (Idexx) and for Map-specific ISMAP02 (VetMAX™-Gold MAP Detection Kit) by direct fecal qPCR. Here, we report individual Map-Ab and Map-excretion profiles from the highest Map prevalent herds out of the 23 dairy herds studied.

The fecal Map excreting and blood Map-Ab profiles revealed great variability among individual animals from the selected MAP infected herds. From the 3452 cows tested, 547 were identified as JD susceptible. Out of these, 169 were considered high shedders since they excreted >106 CFU/g feces, while 378 that excreted 103-5 CFU/g feces and were positive for Map-Ab in one of the samplings were also considered susceptible. Interestingly, after the 2-yr susceptible period, 616 cows showed a latent period greater than 6 yrs during which time fecal shedding was negligible, but fluctuations in blood circulating Map-Ab were detected.

Continuous measurements of faecal Map-Ab excretion and serum Map-Ab level during the longitudinal study allowed us to identify a cohort of cows resistant to progression to the clinical stage of the disease, characterized by long periods of Map incubation and low levels of sporadic faecal excretion. This descriptive analysis highlights the potential for identifying cattle with greater resistant to JD and these animals can then be used to identify genetic markers associated with progression of Map infection and fecal shedding.

206

An integrative Mendelian randomization analysis of genome-wide association and eQTLs studies identified novel genes associated with bovine paratuberculosis susceptibility

Gerard Badia-Bringué[1], Maria Canive[1], Dr Nora Fernández[2], Dr Patricia Vázquez[1], Dr José Luis Lavín[3], Dr José Ramón Bilbao[2], Dr Rosa Casais[4], Cristina Blanco-Vázquez[4], Dr Joseba M. Garrido[1], Dr Ramon A. Juste[1], Dr Marta Alonso-Hearn[1]

[1]Department of Animal Health, NEIKER- Basque Institute for Agricultural Research and Development, Basque Research and Technology Alliance (BRTA), Derio, Spain, [2]Department of Genetics, Physical Anthropology and Animal Physiology, University of the Basque Country (UPV/EHU), Leioa, Spain, [3]Department of Applied Mathematics, NEIKER- Basque Institute for Agricultural Research and Development, Basque Research and Technology Alliance (BRTA), Derio, Spain, [4]SERIDA, Servicio Regional de Investigación y Desarrollo Agroalimentario, Center of Animal Biotechnology, Deva, Spain

Host response and immunology - Poster Pitches, Main Auditorium, Printworks, June 15, 2022, 12:15 PM - 12:30 PM

Genome-wide associations studies (GWAS) have identified genetic variants associated with paratuberculosis (PTB) susceptibility. However, the genes through which these variants exert their effects are unknown and only a few functional mutations or expression quantitative loci (cis-eQTLs) have been characterized. In the current study, the associations between imputed whole-genome sequence data and RNA-Seq data from peripheral blood (PB) and ileocecal valve (ICV) samples of Spanish Holstein cows (N= 16) were analyzed with TensorQTL. This approach allowed the identification of 88 and 38 cis-eQTLs associated with the expression of 90 and 38 genes in PB and ICV samples, respectively (FDR ≤ 0.05). Next, we used summary-based data Mendelian randomization (SMR) to integrate the TensorQTL and GWAS data from a cohort of 813 culled cattle that were classified according to the presence or absence of PTB-associated histopathological lesions in gut tissues. After multiple testing correction, we identified two novel cis-eQTLs tagging the early growth response factor 4 (EGR4) and the bovine neuroblastoma breakpoint family member 6-like protein isoform 2 (MGC134040) that showed pleiotropic associations with the multifocal and diffuse lesions; FDR= 0.002 and FDR= 0.017, respectively. EGR4 is a key regulator of T-cell differentiation and function and MGC134040 is associated with several types of cancer and influences healthy aging. Although no other eQTLs were significant after correction for multiple testing, several genes tagged by the top-associated eQTLs were involved in inflammation, immunity, blood coagulation, and regulation of epithelial cell differentiation, proliferation, and migration. A previously identified cis-eQTL, rs43744169, associated with the up-regulation of the MDS1 and EVI1 complex (MECOM) expression and with increased risk of diffuse lesions was the top eQTL in the PB-SMR database; p[SMR]= 0.0005 and FDR=0.2. Our findings provide a better understanding of the genetic factors influencing PTB susceptibility and highlight genes of importance for follow-up studies.

272

Peripheral cellular immune responses in Mycobacterium avium subspecies paratuberculosis (MAP) infected and vaccinated goats: A comparative study

Dr Shalini Sharma[1,3], Dr Ajay Kumar Gautam[1], Dr Shoor Vir Singh[2], Dr Kundan Kumar Chaubey[2], Dr Saurabh Gupta[2]

[1]Department of Veterinary Physiology and Biochemistry, LalaLajpatRai University ofVeterinary and Animal Sciences, Hisar, Hisar, India, [2]Department of Biotechnology, GLA University, Mathura, Mathura, India, [3]Department of Veterinary Clinical Complex, LalaLajpatRai University of Veterinary and Animal Sciences, Hisar, Hisar, India

Host response and immunology - Poster Pitches, Main Auditorium, Printworks, June 15, 2022, 12:15 PM - 12:30 PM

Background: Th1-mediated immunity play a major role,in protection against mycobacterial infections. Majority of the earlier studies are done in bovine (cattle) and the data on the kinetics of early cellular immune responses following MAP infections in caprine species are very scarce. Furthermore none of the studies have compared the kineticsof early cellular immune responses following experimental infection and vaccination in goats.

Method: Fifteen apparently healthy male kids (3–6 months old) of Barbari breed were included in this experimental study. 5 kids were infected with 'S 5' strain of MAP ("Indian Bison Type"), 5 were vaccinated (Indigenous Vaccine) against MAP infection (Singh et al., 2007) and the remaining 5 kids were uninfected and non-vaccinated controls. Kids were observed for a period of 180 days post exposure (infection and vaccination). Frequencies of CD4 and CD8T cells, plasma IFNγ and TNα and, in vitro cytokine production bypolyclonally stimulated peripheral blood mononuclear cells (PBMCs) from vaccinated, infected and controls were examined.

Results: The frequencies of peripheral CD4 and CD8T cells were comparable in control, infected and vaccinated animals, however around day 49 post-infection, MAP infected animals showed significantly reduced frequencies of CD4 T cells compared to apparently healthy control animals. Plasma TNFα responses in infected animals were significantly reduced compared to controls. In- vitro TNF α production by peripheral blood mononuclear cells (PBMCs) was higher in infected and vaccinated animals at early time points post infection, however the infected animals showed a decreasing trend of cytokine at day 49 to day 180 post-infection.

Conclusions: Peripheral CD4 T cell responses were diminished in MAP infected animals but not in vaccinated animals. It is conceivable that the diminished cellular immune responses may coincide with impairment (immune exhaustion) of CD4T cells that might, in the course of infection, contribute to the progressive nature of caprine paratuberculosis.

188

Validation of a IS900-qPCR assay for the detection of paratuberculosis in faeces according to the OIE "Principles and methods of validation of diagnostic assays for infectious disease".

Dr Simone Russo[1], Dr. Giorgio Galleti[2], Dr Anita Filippi[1], Dr. Norma Arrigoni[1], Dr Chiara Garbarino[1], Dr Matteo Ricchi[1]

[1]IZSLER- Sede di Piacenza - National Reference Laboratory for Paratuberculosis, Via Strada Faggiola, 1 - Podenzano (PC) . Italy, Italy, [2]IZSLER- Sorveglianza Epidemiologica Emilia Romagna (SEER), Via Fiorini 2, Bologna (Italy), Italy

Diagnostics and detection - Poster Pitches (2), Main Auditorium, Printworks, June 15, 2022, 2:15 PM - 2:30 PM

Some authors reported a superior ability of IS900-qPCRs than cultural methods for the detection of Mycobacterium avium subsp. paratuberculosis (MAP) in faecal specimens, but none among these methods has been validated according to the OIE "Principles and methods of validation of diagnostic assays for infectious disease". Our aim was the validation of a IS900-qPCR assay for the detection of MAP DNA in bovine faeces according to this procedure.

Analytical specificity was tested using 47 DNAs from different bacteria, including other than MAP Mycobacteria (exclusivity) and 75 DNAs from MAP (inclusivity). Analytical sensitivity was determined by logit and cLog-Log approaches testing six replicates of negative faeces spiked with MAP decimal dilutions using two DNA extraction methods: semiautomatic with magnetic beads (A) and manual silica column (B). Reproducibility and Repeatability were tested analysing 20 samples in three labs and six samples in one lab during two different working sessions. Two hundred and thirty routine samples also tested with cultural assay (OIE double incubation method) were used to evaluate diagnostic performances by ROC analysis and by bayesian models. Robustness was assessed employing different DNA extraction kits and master mixes.

No amplifications were recorded for no-MAP DNAs while all MAP DNAs tested positive. The analytical sensitivity was 14 or 48 CFU/g and 13or 32 CFU/g for DNA extraction methods A and B and logit or cLog-Log, respectively. Reproducibility and repeatability showed CVs <5%. Area under curve was 0.98 and at 36 Cq cut-off, the diagnostic sensitivity and specificity were 96.1% (IC95% 86.8-99.5%) and 87.6% (IC95% 81.9-92.1%), respectively. Diagnostic sensitivity by bayesian analyses at this cut-off varied from 38.7 to 77.5% while specificity ranged 89.6-94.6%, according to the prevalences and the priors included in the models. No differences were found using different DNA extraction kits and master mixes, underlying the robustness of the assay.

187

Bayesian accuracy estimates of environmental sampling for determining herd paratuberculosis infection status and its association with the within-herd prevalence in Québec dairies

Dr Juan Carlos Arango Sabogal[1], Dr Gilles Fecteau[2], Dr Elizabeth Doré[2], Dr Geneviève Côté[3], Dr Jean-Philippe Roy[2], Dr Vincent Wellemans[2], Dr Sébastien Buczinski[2]

[1]Département de pathologie et microbiologie, Faculté de médecine vétérinaire, Université de Montréal, Saint-hyacinthe, Canada, [2]Département de sciences cliniques, Faculté de médecine vétérinaire, Université de Montréal, Saint-Hyacinthe, Canada, [3]Direction générale des laboratoires et de la santé animale, Ministère de l'Agriculture, des Pêcheries et de l'Alimentation du Québec, Québec, Québec, Canada

Diagnostics and detection - Poster Pitches (2), Main Auditorium, Printworks, June 15, 2022, 2:15 PM - 2:30 PM

Our objectives were to: 1) estimate the sensitivity (Se) and specificity (Sp) of bacterial culture of environmental samples for determining Mycobacterium avium subsp. paratuberculosis (MAP) infection status in Québec dairies, using a Bayesian Latent Class Model (BLCM); and 2) explore the association between the number of positive environmental samples and the individual fecal culture (IFC) apparent and true MAP within-herd prevalence.

Environmental and individual fecal samples were collected from 87 commercial dairy herds participating in previous research projects. Samples were cultured using the MGIT Para TB culture liquid media and the BACTEC MGIT 960 system. The Se and Sp of environmental sampling were estimated using a one-test-one-population BLCM. Herds were considered positive for environmental sampling if at least one out of the six samples collected was positive. The apparent and true IFC within-herd MAP prevalence estimates for each herd were obtained using a two-stage cluster BLCM. The association between the apparent and true within-herd MAP prevalence results, and the number of positive environmental samples was assessed using a zero-inflated negative binomial (ZINB) model. In all BLCMs, median posterior estimates and 95% Bayesian credible intervals (BCI) were obtained with OpenBUGS statistical freeware.

Se and Sp of environmental sampling were 43.7% (95% BCI: 32.5-55.5) and 96.2% (95% BCI: 84.2-99.8), respectively. The number of positive environmental samples increased with the apparent and true MAP within-herd prevalence. The true prevalence was higher than the apparent prevalence for a given number of positive environmental samples.

Despite its imperfect accuracy, environmental sampling is an inexpensive and non-invasive sampling method to determine MAP infection status in tie-stall herds that can be used as a proxy to estimate the true within-herd prevalence. The absence of positive environmental samples in a single sampling visit was likely an indicator of a very low within-herd prevalence rather than being MAP exempt.

289

Identification of Mycobacterium avium subsp. paratuberculosis using a single-chain antibody targeting the cell wall lipopentapeptide L5P

Ph.d. Horacio Bach[1], MSc Yael Slavin[1], MSc Ariana González[1], Ph.D. Christelle Ganneau[2], Ph.D. Sylvie Bay[2], Ph.D. Franck Biet[3]

[1]University of British Columbia, Vancouver, Canada, [2]2Institut Pasteur, Unité de Chimie des Biomolécules, CNRS UMR 3523, 75724 Paris Cedex 15, France, [3]INRA, Université de Tours, UMR1282, Nouzilly, France

Diagnostics and detection - Poster Pitches (2), Main Auditorium, Printworks, June 15, 2022, 2:15 PM - 2:30 PM

Introduction: The detection of Mycobacterium avium subsp. paratuberculosis (MAP) in biological samples has been hampered because of the lack of reliable antibodies able to specifically recognize components of its cell wall. The subspecies MAP produces specific lipopentapeptide L5P or lipotripeptide L3P on the cell wall surface, which has not been yet found so far in other mycobacterial species.

Purpose: Development of single-chain antibodies (scFvs) for a rapid identification of MAP targeting the L5P antigen.

Methodology: Phage display technology used a phage library of scFv antibodies to screen antibodies against the cell wall L5P extracted from MAP strain K-10. The scFv termed L7 was produced as a recombinant protein in E. coli and purified by affinity chromatography. The portion of the L5P molecule that binds the antibody was determined by exposing the antibody to seven derivatives of the L5P using fluorescent microscopy. Furthermore, the cross-reactivity of the antibody with other mycobacterial species, such as M. abscessus, M. avium subsp. avium, M. bovis, and M. smegmatis were assessed by immunostaining.

Results: L7 was successfully produced as a soluble recombinant protein in E. coli. No cross-reaction was observed when L7 was exposed to other mycobacterial species, suggesting that L7 is specific to MAP. Furthermore, binding analysis of various L5P analogs showed that L7 recognizes only the pentapeptide portion of the molecule containing the amino acid sequence D-Phe-L-NMe-Val-L-Ile-L-Phe-L-Ala, and not the lipidic moiety or the tripeptide sequence of the L3P present on the sheep type MAP cell wall. Lastly, the antibody was able to identify MAP in milk, colostrum, and biopsies by immunofluorescence.

271

Evaluating the performance of various protocols for the diagnostic performance in paratuberculosis IS900 PCR

Dr Mukta Jain[1], Ms Harshita Gandhi[1], Ms Simpi Mourya[1], Dr Amit Kumar[2], Dr Gajendra Aseri[1], Dr Deepansh Sharma[1], Dr Jagdip Singh[1]

[1]Amity Center For Mycobacterial Disease Research, Jaipur Rajasthan, India, Jaipur , India, [2]E-YUVA Centre, Panjab University, Chandigarh, Chandigarh , India

Diagnostics and detection - Poster Pitches (2), Main Auditorium, Printworks, June 15, 2022, 2:15 PM - 2:30 PM

Paratuberculosis is a prevalent infectious disease of domestic ruminants resulting in huge economic loss. Zoonotic evidences are also mounting. Therefore, control of paratuberculosis is a priority. Vaccine is available but use is limited due lack of DIVA. Control is largely dependent on Test and Cull/Segregate policy. Hence, it is critical to assure on the infection status before making the decision. Commercial kits are available; however, high cost of imported kits limits their use especially in resource poor countries. Therefore, inhouse and native alternatives are needed. PCR based diagnosis is definitive tool to confirm infection. In the present study five DNA isolation protocols were evaluated six (Method I: magnetic nanoparticles based, Method II: biochemical protocol, Method III: physiochemical protocol, Method IV: filter paper protocol, Method V: Chemical method and Method VI: physical protocol). Further, four types of PCR master mixes were also tested (MM I: Qiagen based, MM II, Thermo based, MM III, Genei based with additives and MM IV Genei based with added MgCl2). On pilot scale fecal samples (100) physical method of DNA isolation (Method VI) in conjunction with Master Mix II (MM II) showed highest sensitivity and was then used to test large panel of samples. This DNA isolation method (Method VI) is based of concentration of cells followed by heat-based lysis. Of the total 521 animals tested, 120 (23%) were found positive using IS900 PCR. Species wise disease was more frequent in cattle (32.8%) followed by goat (20%), buffalo (19.3%) and sheep (11.3%). In conclusion, findings of the present study report the usefulness of the in-house fecal IS900 PCR method in field studies. Use of additives to remove PCR inhibitors may further increase the diagnostic sensitivity. High prevalence of paratuberculosis needs immediate implementation of the control programs.

186

Herd Environmental Sampling for detection of Mycobacterium avium subspecies paratuberculosis in pasture-based dairy herds – preliminary results

Miss Niamh Field[1], Dr. John Mee[1], Dr. Conor McAloon[2]
[1]Teagasc, Fermoy, Ireland, [2]University College Dublin, Dublin 4, Ireland

Diagnostics and detection - Poster Pitches (2), Main Auditorium, Printworks, June 15, 2022, 2:15 PM - 2:30 PM

The objective of this study was to determine the herd sensitivity (HSe) of herd environmental sampling (HES) for the detection of Mycobacterium avium subspecies paratuberculosis (MAP) in pasture-based Irish dairy herds.

Materials and methods: In a two-year study (2019-2021), 122 commercial dairy herds were recruited from the Irish Johne's Control Programme (IJCP). Each herd was visited once, when cows were housed for the winter. Two herd-level tests for MAP were conducted at each visit: whole-herd serum ELISA with confirmatory faecal PCR of seropositive animals (sELISA + fPCR), and HES. The sampling protocol for HES consisted of six composite environmental samples obtained from areas of manure concentration: slurry storage, cow housing, the collecting yard and calving pens. Blood samples were tested using the IDEXX MAP ELISA kit and HES samples were tested by PCR using the Indical Bactotype MAP PCR kit. Individual faecal samples were tested with various different PCR kits.

Results: There were results on both test methods for samples from 97 herds. Seventeen herds (17.5%) were identified as infected by any test (three positive on both tests, 12 positive on sELISA + fPCR only and two positive on HES only). Fifteen herds were identified as infected using sELISA + fPCR. Of these, three herds were also identified as infected by HES (20% relative HSe). Five herds were identified as infected by HES, but two of these herds were negative on sELISA + fPCR. The majority of herds (82%) tested negative on both tests.

Conclusions: In these preliminary results, HES detected fewer infected dairy herds than the current standard method of testing herds for MAP in Ireland (sELISA + fPCR). Further analysis is required to accurately estimate the HSe and HSp of HES to determine if it is suitable for use as an alternative or adjunct herd-level test.

320

Invited Speaker Epidemiology and Economics

Emeritus Professor Frank Griffin

Epidemiology and Economics - Invited Speaker, Main Auditorium, Printworks, June 16, 2022, 8:30 AM - 9:15 AM

Five learning points

1. Complex diseases caused by mycobacterial pathogens will not be resolved by the application of singular simple remedies.
2. The economic impacts of infectious diseases are cumulative and should incorporate the added management costs, combined with overall production losses.
3. Effective control of chronic infectious disease in livestock demand co-incident remediation strategies, supported fully by producers, scientists, veterinarians and regulators.
4. For cost-effective control of chronic infectious disease in farm animal, equal contributions are made; through altered management (farmers), test performance (laboratories), epidemiology (veterinarians) and remedial interventions (regulators).
5. Composite strategies for disease control are synergistic, where the sum of the parts are greater than the whole.

306

Dam and environmental transmission routes on Mycobacterium avium paratuberculosis infection in calves

Miss Erica Nunney[1], Dr Matteo Crotta[1], Ms Karen Bond[2], Professor Martin Green[3], Dr Steven Van Winden[1], Professor Javier Guitian[1]

[1]Royal Veterinary College, Hatfield, United Kingdom, [2]National Milk Records Group, Chippenham, United Kingdom, [3]Univerisity of Nottingham, Sutton Bonnington, United Kingdom

Epidemiology and Economics - AM Session (1), Main Auditorium, Printworks,
June 16, 2022, 9:15 AM - 10:30 AM

Cow to calf transmission of Mycobacterium avium subsp. paratuberculosis (MAP) can occur in utero, via milk and colostrum or faecal-orally. Understanding of the relative role of the different transmission routes to calves is important in formulating control recommendations. Our aim in this longitudinal study was to measure the association between indicators of transmission via the dam and the environment on a calf testing positive for MAP.

The study population comprised of 440 UK dairy calves from 6 herds enrolled between 2012-2013. These calves were followed up from birth until 2021, when most of them had completed their productive lives. At birth, individual calf data were captured. During follow-up, individuals entering the milk herd were quarterly tested for the presence of MAP using milk ELISA testing. Cox regression models were used to measure the association between transmission routes via the dam (in-utero and colostrum) and via environmental (time spent and cleanliness of yard) and time till first detection of MAP infection.

The hazard ratio for calves born from positive dams was 1.99 (95% CI: 1.2-3.2). After partial adjustment for confounding, to estimate in utero route, calves born to positive dams were still associated with an increased probability of having a JD positive test result (HR: 2.36; 95%CI: 1.2-4.7). Examining cows unlikely to being infected with MAP via the in-utero or colostrum route, the hazard ratio for calves that had been kept longer in yards with worst cleanliness score was 4.51 (95% CI: 1.15-17.62).

Our results, which have been partially controlled for confounding, suggest the in-utero route plays an important role in MAP transmission and provide an estimation of the relative contribution of dam-to-calf vs. environmental transmission in endemic herds, which is important to refine farm-level control strategies.

195

Economic premiums associated with Mycobacterium avium subspecies paratuberculosis-negative replacement purchases in major dairy producing regions.

Dr Philip Rasmussen[1], Herman W. Barkema[2], Eugene Beaulieu[2], Steve Mason[2], David C. Hall[2]
[1]University Of Zurich, Zurich, Switzerland, [2]University of Calgary, Calgary, Canada

Epidemiology and Economics - AM Session (1), Main Auditorium, Printworks, June 16, 2022, 9:15 AM - 10:30 AM

The economic losses due to Johne's disease (JD) in dairy cattle herds and the benefits and costs of various potential control practices have been estimated before. However, little is known about the economic value of purchasing MAP-negative dairy replacements in major dairy-producing regions. This study used Markov Chain Monte Carlo (MCMC) simulation techniques to compare two sets of MAP-negative and MAP-positive herds across a comprehensive selection of regions: herds purchasing MAP-negative replacement animals and herds purchasing replacement animals with unknown MAP infection status. The economic benefits in terms of improved production, longevity, and salvage value per MAP-negative replacement purchased were then estimated over a 10-year horizon, and the additional value of MAP-negative replacements when compared to unknown status replacements were calculated as a percentage premium of the average aggregated dairy replacement price in each region. Due to the unique market conditions that arise as a result of supply management in Canada, this comparison was repeated a second time for Canada with production benefits instead measured as decreased variable costs (labour, veterinary fees, bedding, feed, etc.). An average benefit of US$76 per MAP-negative replacement purchase was estimated in major dairy-producing regions, equivalent to a premium of 13%, with higher premiums in regions characterized by below-average replacement prices and on-average farm-gate prices. For Canadian dairy herds, it was estimated that the economic benefits range from US$80 to US$139 per MAP-negative replacement, or 8 to 14% of average aggregate replacement prices. It was also estimated that the greatest benefits from MAP-negative replacement purchases are associated with MAP-negative herds that successfully remain uninfected. This research suggests that MAP-negative replacements have significant additional value, and that given the dissemination of accurate pricing information and an organized market, this value could be captured, at least in part, by herds that successfully control JD.

216

Effects of Silirum® based vaccination programs on Map faecal shedding and serological response in seven French dairy herds.

Pr Fabien Corbiere[1], Dorra Guellouz[1], Julie Cournet[1], Christian Tasca[1], Emma Dubaux[2], Laurent Foures[2], Pr Gilles Foucras[1]

[1]INRAE & Toulouse National Veterinary School, Toulouse, France, [2]GDS Meuse, Verdun, France

Epidemiology and Economics - AM Session (1), Main Auditorium, Printworks, June 16, 2022, 9:15 AM - 10:30 AM

The objective of this study was to evaluate the effects of Silirum® based vaccination programs on Map faecal shedding and serological status in French dairy herds infected with paratuberculosis. To this end, the serological status (ELISA) and faecal shedding (qPCR) of 4- to 6-year-old vaccinated cows (n=372) were assessed twice to 3 times over a 3-year period in 7 infected herds in the Meuse department, France. The median age at vaccination was 4.4 months (interquartile range: 2.9-7.2 months). Within each herd, cows from the last non-vaccinated birth cohort (n= 265) were used as controls. Data were analyzed using mixed general linear regression models adjusted on days in milk and age at sampling, with herd and cow random effects. Only 23.7% of the vaccinated cows were positive on serum ELISA, with cows vaccinated before the age of 6 months yielding significantly fewer positive results. Overall, 38.3% of vaccinated and 29.8% of non-vaccinated cows were positive on faecal qPCR, with strong differences between herds, but high shedding levels were observed in only 6.5% of cows. Compared to non-vaccinated seronegative cows, the probability of Map shedding was significantly lower only in cows vaccinated before 4 months of age (Odd's Ratio = 0.54, 95 % confidence interval: 0.33 – 0.90, p=0.016), while non-vaccinated seropositive cows were at higher risk (OR = 3.95, 95%CI: 1.42 – 10.97, p=0.008). The amount of Map shed was the highest in non-vaccinated seropositive cows while no difference was evidenced between non-vaccinated seronegative cows and vaccinated ones, whatever their age at vaccination. Based on these results, a beneficial effect of vaccination on Map faecal shedding may exist in cows that have been vaccinated before 4 months of age. Moreover, the variability of serum ELISA response in vaccinated cows remains to be investigated.

199

Insights on paratuberculosis in cattle at a regional scale using data-driven modeling and inference

Dr Gaël Beaunée[1], Dr Pauline Ezanno[1], Mr Alain Joly[2], Dr Pierre Nicolas[3], Dr Elisabeta Vergu[3]
[1]*BIOEPAR, INRAE, Oniris, 44307 Nantes, France,* [2]*GDS Bretagne, 56003 Vannes, France,* [3]*NRAE, Université Paris-Saclay, MaIAGE, 78350 Jouy-en-Josas, France*

Epidemiology and Economics - AM Session (1), Main Auditorium, Printworks, June 16, 2022, 9:15 AM - 10:30 AM

A better understanding of the mechanisms underlying the spread of bovine paratuberculosis on a large scale is a key point for the implementation of effective control strategies and the development of accurate decision support tools. Modern inference methods now allow acquiring new knowledge from observed data when combined with mechanistic modelling.
Our objective was to estimate key parameters of the spread of Mycobacterium avium subsp. paratuberculosis (Map) between cattle herds from longitudinal and spatial data collected in Brittany (Western France) consisting of serological tests conducted in 2,278 herds sampled between 2007 and 2013. We estimated the proportion of herds infected in 2005, the within-herd prevalence distribution, the probability of purchasing infected cattle from outside the metapopulation considered, the indirect transmission rate within infected herds, and the mean sensitivity of the diagnostic test.
We used a detailed stochastic mechanistic model of Map spread within and among dairy herds through animal trade movements at the regional scale (12,857 dairy herds). Inference was performed with a Monte-Carlo approximation of a composite likelihood coupled to a numerical optimization algorithm. The validity of the inference method was verified on simulated data.
The estimation results indicated a situation in 2005 with a very large proportion of infected herds in the metapopulation (> 80%) but with a low average within-herd prevalence (≈0.16). They also confirmed the low diagnostic test sensitivity, with an average estimate of 0.21. The risk of purchasing an infected animal from outside the region was estimated constant over time and moderate (≈0.13).
Estimations of such previously unknown parameters provide new insights on the dynamics of bovine paratuberculosis in Western France, bringing more robust predictions accounting for regional specificities in terms of contact network and farms' characteristics. The inference framework proposed here can be applied easily to other infected areas where similar datasets are available.

232

Success of Paratuberculosis Control is associated with Calf Mortality Rate in Dairy Herds

Dr. Anne Klassen[1], Dr. Stefanie Söllner-Donat[1], Dr. Esra Einax[1], Dr. Karsten Donat[2]

[1]Animal Health Service, Thuringian Animal Diseases Fund, Jena, Germany, [2]Clinic for Obstetrics, Gynaecology and Andrology of Large and Small Animals with Veterinary Ambulance, Justus Liebig University Giessen, Giessen, Germany

Epidemiology and Economics - AM Session (1), Main Auditorium, Printworks, June 16, 2022, 9:15 AM - 10:30 AM

Calf health is assumed to be associated with on-farm paratuberculosis control activities like improved hygiene of calving and calf rearing. This study aimed at determining this association in dairy herds enrolled in a voluntary regional paratuberculosis control programme in Germany.

Forty-one paratuberculosis affected dairy herds were monitored for MAP shedding cows and calf mortality from 2008 to 2017. Annual cumulative incidence of MAP shedders was calculated from the annual individual faecal culture testing results. Herds were classified depending on whether they successfully reduced MAP shedders' incidence to a non-detectable level during this period (certified as paratuberculosis non-suspect herds, n=13) or not (n=28). Annual calf mortality rate was calculated using the official data base for animal movement. Farm-specific time series of these data were adjusted by the year of enrolling in the control program. In 2017 a survey with detailed questions regarding hygiene of calving, nutrition and housing of calves was conducted in each farm. A generalized mixed linear model with repeated measured values was used to investigate the association between herd groups and calf mortality rate.

Herds that achieved the "non-suspect" status had a lower calf mortality rate compared with farms which did not reach that level of incidence reduction. Two years after implementation of control measures this difference was most distinct. Over time, mortality rate reduced in both groups of farms.

The results show for the first time that establishing a paratuberculosis control programme positively affects calf mortality rate and that farms which successfully reduced the incidence of MAP shedders to a non-detectable level have a lower calf mortality rate than farms with less reduction of MAP shedders' incidence.

207

Mycobacterium Avium subspecies Paratuberculosis (MAP) Elisa tests on milk to predict future health and culling in a dairy herd

Mr Richard Sibley[1], Mr Peter Orpin[1]
[1]West Ridge Vets, Witheridge, United Kingdom

Epidemiology and Economics - AM Session (2), Main Auditorium, Printworks, June 16, 2022, 11:00 AM - 12:15 PM

This study investigates the use of serial milk MAP elisa tests to predict health, productivity and emergency culling of adult dairy cows.
The study involved a large dairy herd of over 900 adult cows using milk MAP elisa tests on all milking cows every three months as part of a Johnes Control programme. Over 3000 adult cows that had been culled from the herd over a period of ten years, and which had milk elisa test results for their entire adult life were studied to determine if test results in the could be used to create a robust plan to manage test positive cows to predict and prevent health problems. Udder health, lifetime milk yield, fertility and emergency culling (as defined by being culled within 100 days of calving) were used to define health and productivity.
Cows that had just one positive MAP elisa test in their lifetime had similar health and productivity to cows that had never had a positive test. Cows that had at least 2 consecutive positive tests had significantly higher somatic cell counts and higher emergency culling rates (34% of these cows were culled within 100 days of calving). Cows with at least four positive MAP elisa tests, with rising titres at each consecutive test had a very high emergency culling rate, such that 46% of these cows were culled within 100 days of calving for severe health issues. There was little difference in milk yields between positive and negative test cows.
Cows with repeated positive MAP test results have high risks of health problems that lead to emergency culling, with the consequent major economic losses incurred when culling lactating cattle with no salvage value. Rising titres of MAP antibodies as measured by milk elisa tests indicate imminent severe health problems with poor prognosis.

228

Exposure to Mycobacterium avium subsp. paratuberculosis in Alpine pastures (Northern Italy): evaluation of cattle and red deer (Cervus elaphus) contribution through environmental faecal samples.

Dr Anita Filippi[1], Dr Camilla Luzzago[2], Dr Matteo Nava[2], Dr Alessandro Forti[2], Dr Simone Russo[1], Dr Filippo Barsi[1], Dr Luca Corlatti[3], Dr Luca Pedrotti[3], Dr Alessandro Bianchi[4], Dr Matteo Ricchi[1], Dr Matteo Ricchi, Dr Norma Arrigoni[1], Dr Chiara Garbarino[1]

[1]Istituto Zooprofilattico Sperimentale della Lombardia e dell'Emilia Romagna, Territorial Unit of Piacenza, Piacenza, Italy, [2]Department of Veterinary Medicine, Università degli Studi di Milano, Milano, Italy, [3]Stelvio National Park, Bormio, Italy, [4]Istituto Zooprofilattico Sperimentale della Lombardia e dell'Emilia Romagna, Territorial unit of Sondrio, Sondrio, Italy

Epidemiology and Economics - AM Session (2), Main Auditorium, Printworks, June 16, 2022, 11:00 AM - 12:15 PM

In the Lombardy area of the Stelvio National Park (Northern Italy) a 20% prevalence of MAP was reported in wild red deer population (2011-2015). To assess the epidemiological role of pasture contamination and the different contribution of bovine and red deer in this area, we tested the presence of MAP during the grazing period (August 2020) by sampling faeces of both species.

Twenty-one areas were selected, either shared or not shared by deer and bovine ; deer abundance was approximated with density estimates of faecal pellets using a distance sampling approach.
Overall, 757 faecal samples, 470 from deer and 287 from bovine, were collected and analysed by qPCR for MAP detection. The state of faeces (fresh, semidry, dry), faecal sample mass and geographical coordinates were recorded. Positive samples were cultured for MAP, genotyping using MIRU-VNTR typing and Short Sequence repeats; pathogen shedding was quantified by digital PCR.

A total of 19 (2.5%) samples, 16 from deer and 3 from bovine, were positive to qPCR assay. Six strains were isolated among qPCR positive deer samples, from the area where the highest number of positive results was recorded (8 in deer and 1 in cattle). In total, nine areas out of 21, shared or not shared, resulted positive. All positive samples came from fresh or semidry faeces. Digital PCR showed a low level of MAP faeces contamination (< 1 copy/µL). All isolates showed the same INMV1 genotype, already described in the area.

Our results show a wide but low contamination of Alpine pastures revealing dissemination of MAP not only in wild deer population but also in cattle grazing into the park. Additional sampling, selecting fresh faeces, is planned to obtain more comprehensive and reliable results that will serve for adopting control strategies to avoid interspecific transmission.

225

Factors associated with the spread of paratuberculosis in Hungarian dairy cattle herds

Dr Viktor Jurkovich[1], Dr Mikolt Bakony[1], Dr Barbara Vass-Bognár[1]
[1]*University of Veterinary Medicine, Budapest, Hungary*

Epidemiology and Economics - AM Session (2), Main Auditorium, Printworks, June 16, 2022, 11:00 AM - 12:15 PM

The aim of our study was to identify the hazards associated with the spread of paratuberculosis (PTB) during the period of calving and calf rearing in dairy herds. Secondly, we wanted to select some of the identified hazards whose changes can ease the eradication process.
Our studies were carried out in 27 Hungarian large-scale dairy farms, using a questionnaire. The average herd size was 911 (min. 152; max. 2200; 24,599 cows in total). Based on the literature, eleven different management aspects of our questionnaire were considered influential in the spread of MAP from adult cows to calves. Farms were evaluated, and results were dichotomized based on whether the given management practice was advantageous (1) in controlling the spread of MAP or disadvantageous (0). By summing up the obtained values, an overall score was achieved that would rank the farms based on their “awareness” in MAP control. The random forest classification method was used to identify variables that were most decisive in the magnitude of apparent MAP seropositivity.
The average apparent animal-level PTB prevalence was 8.3% (min. 2.0, max. 19.5) in the visited herds. Taking the calf away immediately after birth was associated with an average 52% decrease in the odds of seropositivity as compared to the longer time spent with the dam. The destroying of MAP-infected milk was associated with a 43% decrease in the odds of seropositivity as compared to farms where MAP-infected milk was not destroyed. The feeding milk replacer solely after birth was associated with a 36% decrease in the odds of seropositivity as compared to feeding also milk. Other factors, such as the distance of ill and calving cows, drinking of treated milk from sick cows and individual or group calving are also influential.
Our results showed that calf management is key in controlling paratuberculosis.
Viktor Jurkovich was supported by the Ministry of Innovation and Technology and the National Research, Development, and Innovation Fund (grant No.: MEC_21_R_140844).

236

Multi-institutional investigation for determining the status of Johne's disease in the suspected and apparently healthy population of domestic livestock species: A Pan-India Study

Dr Shoor Vir Singh[1,6], Dr D.S. Chauhan[2], Dr J.S. Sohal[3], Dr K. Gunaseelan[4], Dr A.V. Singh[2], Dr K Porteen[5], Dr Kundan Kumar Chaubey[1,6], Dr Saurabh Gupta[1,6], Dr P.K. Singh[1]

[1]*Department of Biotechnology, GLA University, Mathura, Mathura, India,* [2]*ICMR-National JALMA Institute for Leprosy and Other Mycobacterial Diseases, Agra, Agra, India,* [3]*Amity Institute of Microbial Technology, Amity Center for Mycobacterial Disease Research, Amity University Rajasthan, Jaipur, Jaipur, India,* [4]*Tamil Nadu Veterinary and Animal Sciences University (TANUVAS), Madras Veterinary College, Chennai, Chennai, India,* [5]*Department of Veterinary Public Health And Epidemiology, Madras Veterinary College, Chennai, Chennai, India,* [6]*ICAR-Central Institute for Research on Goats, Makhdoom, Mathura, India*

Epidemiology and Economics - AM Session (2), Main Auditorium, Printworks, June 16, 2022, 11:00 AM - 12:15 PM

Background: Johne's disease (JD) is a major health problem and endemic in the domestic livestock population world-wide including India. In past >50 years, different workers reported highly variable estimates of JD in the country. Therefore present study involving five institutions was planned to investigate status of JD in the domestic livestock population located in different geographical regions (North, West, East and South) using multiple diagnostics.

Methods: Screening of livestock species was conducted from 2015 to 2018 in different regions (2 centres from Northern and one each from Southern, Eastern and Western) of country for the presence of Johne's disease using microscopy, PCR and ELISA. A total of 749 animals (clinically suspected: 515 and apparently healthy: 234) were screened by fecal microscopy and PCR. Additionally, 4850 serum were also screened randomly irrespective of clinical symptoms using indigenous ELISA. Results were stratified and analysed geographically.

Results: Of 749 animals screened, 23.4 and 24.8% animals were positive by smear microscopy and PCR, respectively. MAP positivity among suspected animals was 29.3 and 32.8% by microscopy and PCR; whereas it was lower among apparently healthy animals (microscopy: 10.3%; PCR: 7.3%). Considering PCR as confirmatory test, prevalence of JD among suspected animals was highest in Western region (52.3%), followed by Northern (31.1%), Southern (23.4%), and Eastern (15.0%) regions. A high proportion of apparently healthy animals from Northern region (35.1%) were positive by IS900 PCR, and was lower in other regions. Sero-prevalence of JD by indigenous ELISA was 54.2% (2630/4850). Geographically, sero-positivity was highest (67.3%) in Southern region followed by Western (64.1%), Northern (51.6%) and Eastern (33.3%) regions.

Conclusions: Study reported high bio-load of JD in the domestic livestock population of the country though percent prevalence was variable, test and region-wise, which warrants immediate control measures by the government.

318

Epidemiology of Johne's disease in the New Zealand farmed deer industry 1980-2020

Dr Solis Norton[1], Dr Marian Price-Carter[2], Emeritus Professor Frank Griffin
[1]*Deerpro, Dunedin, New Zealand,* [2]*AgResearch, Palmerston Norton, New Zealand*

Epidemiology and Economics - AM Session (2), Main Auditorium, Printworks, June 16, 2022, 11:00 AM - 12:15 PM

The New Zealand (NZ) farmed deer industry was established in the 1980s and contains around 1 million deer across 1200 farms. Its index case of Johne's disease (JD) occurred in 1986. The disease spread rapidly. On-farm, outbreaks could be severe with death rates up to 15% reported in young deer. By 2011 there was an estimated 66% herd prevalence.

Whole Genome Sequencing (WGS) of 38 isolates of the causal organism, Mycobacterium avium subspecies paratuberculosis (MAP), cultured from farmed deer from 8 regions throughout NZ between 2010 and 2014, suggested three recent introductions. The introduced types had more in common with genomes from overseas ruminant isolates than with those from 29 geographically diverse NZ cattle isolates with prevalent VNTR/SSR types. WGS of 4 deer isolates from late in 2020 indicate that at least 2 of these types are persisting in NZ deer herds.

A national slaughterhouse surveillance database and control programme were established in 2006 to monitor signs of the disease recorded during routine deer carcass inspection and assist in its control.

Based on 5.2 million slaughterhouse surveillance records, evidence of disease in the total population of slaughtered young deer has halved from its peak of 4-6% per month in 2008-2011 to 2-3% per month in 2020-2021. In 2013-14, the association between evidence of JD and lighter carcass weight in young deer peaked at 5kg. Gradually declining thereafter, it had halved by 2020-21. Anecdotal evidence from farmers and veterinarians is that severe outbreaks on-farm have now virtually disappeared although most farms remain infected.

In conclusion, from 1986 to 2021 Johne's disease in the New Zealand farmed deer industry has traversed from epidemic to chronic phase. Genetic evidence suggests this is not due to changes in MAP types. Possibly the farmed deer population have become more resilient to infection.

Seroprevalence of Mycobacterium avium subsp. paratuberculosis in Austrian goat flocks

Miss Juliane Schrott[1], Dr. Eva Sodoma[2], Dr. Michael Dünser[1,2], Dr. Alexander Tichy[3], Dr. Johannes Lorenz Khol[4]
[1]Austrian Agency for Health and Food Safety (AGES), Institute for Veterinary Disease Control Innsbruck, , Austria, [2]Austrian Agency for Health and Food Safety (AGES), Institute for Veterinary Disease Control Linz, , Austria, [3]Platform for Bioinformatics and Biostatistics, Department of Biomedical Sciences, University of Veterinary Medicine Vienna, , Austria, [4]University Clinic for Ruminants, Department for Farm Animals and Veterinary Public Health, University of Veterinary Medicine Vienna, , Austria

Epidemiology and Economics - Poster Pitches, June 16, 2022, 12:15 PM - 12:30 PM

This study gains first information about the current epidemiological situation of infection with Mycobacterium avium subsp. paratuberculosis (MAP) in Austrian goat flocks. As goat production in Austria is a growing sector, there are rising concerns regarding the spread of MAP infection and the linked economic impact. The aim of our study was to estimate the MAP herd level seroprevalence in Austrian goat flocks and to identify possible associated risk factors.
Therefore, serum samples from 6,434 goats, originating from 638 flocks, were tested for the presence of MAP specific antibodies, using a commercial ELISA. Flocks were selected randomly, considering a proportional provincial distribution, with the sample size of each flock depending on the flock size. Furthermore, larger flocks, dairy goat herds and farms with increased animal trading were assigned with a higher selection probability. Productive parameters were gained from the national animal health information system. Risk factors associated with MAP seroprevalence were estimated using chi-square test and binary logistic regression.
The investigation revealed an animal level apparent MAP seroprevalence of 2 % (a total of 126 goats were seropositive). In 71 flocks, at least one seropositive animal was detected, leading to a herd level apparent MAP seroprevalence of 11 %. Logistic regression showed significant association between MAP seropositivity and flock size. Seroprevalence was also significantly higher in dairy goat herds and farms with animal trading. Furthermore, significant differences between regions were observed.
The results indicate a moderate MAP infection rate in goat flocks in Austria. However, we notice a mismatch between the numbers of samples from goats analysed relating to the compulsory paratuberculosis control program and the investigated MAP herd level seroprevalence. To contain a future spread of MAP among goat flocks, herd-level investigations and prevention measures should be established.

273

Understanding the antibacterial mechanisms of copper ion treatment on Mycobacterium avium subsp. paratuberculosis

Mr Carlos Tejeda[1], Mss Marcela Villegas[1], Dr Pamela Steuer[1], Dr Miguel Salgado[1]

[1]Instituto de Medicina Preventiva Veterinaria, Facultad De Ciencias Veterinarias, Universidad Austral De Chile, Valdivia, Chile

Epidemiology and Economics - Poster Pitches, June 16, 2022, 12:15 PM - 12:30 PM

Introduction: Copper and its alloys are natural and very well-proven antimicrobial materials. The mechanisms of action through which copper is highly effective to have been described at the molecular and cellular levels. However, both the design of the studies carried out and the nature of the microorganisms studied have meant that this research has been of limited scope. In the present study, we examined the action mechanisms of a copper ion treatment on the integrity of Mycobacterium avium subsp. paratuberculosis (MAP), a highly resistant animal pathogen.

Material and Methods: To investigate the mechanisms that would explain how copper ions exert effective control on MAP, a study was proposed that evaluated the integrity of the nucleic acids (Comet assay), proteins (ROS and oxidation proteins), and cell wall (lipid oxidation) of this pathogen, as response variables against this treatment.

Results: The copper ion treatment applied to MAP cells resulted in nucleic acid degradation and disintegration, increased ROS production, and protein alteration. However, it had no effect on the integrity of the MAP cell wall, which offers a possible scientific explanation for the observed tolerance of this pathogen to this copper-based treatment.

Conclusion: This new evidence about the observed tolerance in the MAP cell wall against the copper ion may help us to understand how we can improve the proposed copper-based treatment, and finally achieve a totally effective alternative to control MAP.

254

Molecular and serological survey of Mycobacterium avium subspecies paratuberculosis in Ugandan cattle

Dr. Judah Ssekitoleko[1,6], Prof Lonzy Ojok[1], Dr Elsagad Eltyayeb[2], Prof Ahmed Amanzada[3], Dr Kamal Eltom[4], Dr Ahmed Abd El Wahed[5], Dr Julius Okuni[1]

[1]Department of Veterinary Pharmacy, Clinical and Comparative Medicine, Makerere University, Kampala, Uganda, [2]Ibn Sina Specialised Independent Hospital, Alneelain University, Khartoum, Sudan, [3]Department of Gastroenterology and Gastrointestinal Oncology, University Medical Centre Goettingen, Georg-August University Goettingen (UMG), Goettingen, Germany, [4]Unit of Animal Health and Safety of Animal Products Institute for Studies and Promotion of Animal Exports University of Khartoum , Khartoum, Sudan, [5]Institute of Animal Hygiene and Veterinary Public Health University of Leipzig, An den Tierkliniken 43, D-04103 Leipzig , Leipzig, Germany, [6]National Agricultural Research Organisation, Entebbe, Uganda

Epidemiology and Economics - Poster Pitches, June 16, 2022, 12:15 PM - 12:30 PM

Mycobacterium avium subspecies paratuberculosis (MAP) infection in cattle occurs worldwide but in many countries, there is still limited information about its occurrences within herds at local and national levels. Although a few studies have reported the occurrence of MAP infection in Ugandan cattle, they were limited to only the central region.The aim of this report is to describe the molecular and serological prevalence of MAP and epidemiological characteristics in herds of cattle in South Western Uganda. A multistage cluster Sampling was carried out in six districts of Uganda, namely Kabale, Mbarara, Kiruhura, Bushenyi, Ntoroko and Ntungamo. Serum and faecal samples were obtained from one thousand eight hundred fourteen (1814) cattle from 93 herds and tested using IDEXX ELISA and Recombinase Polymerase Amplification (RPA) targeting IS900 of MAP respectively. A questionnaire was administered to collect epidemiological information from the cattle owners. The overall cow prevalence was 3.19% by serum-ELISA and 2.7% using RPA-Faecal testing; with a herd prevalence of 45.8% for herds which had at least one cow testing positive. The prevalence of MAP infection was lowest in Kiruhura (1.6% using Seum-ELISA and 0.7% using Faecal-RPA) and highest in Ntoroko (5.6% using serum-ELISA, 5.6% using Faecal-RPA). The average district prevalence was 3.4 ± 1.25 and 2.9 ± 1.65 using serum-ELISA and Faecal – RPA respectively. Within herd prevalence ranged from 1.6 to 7.3% by serum-ELISA. The prevalence of MAP infection in the six districts remains low as it has been found in other districts in Uganda using both serological and DNA testing however, the there is cause to worry since the herd level prevalence is almost 50% and within herd prevalence of some herds were also reaching about 25%. The cattle were of mixed breeds kept on small and medium farms under communal grazing and paddocks.

241

Johne's disease in New Zealand; the study of a milk antibody dataset

Miss Rebecca White, Dr Shona Pryor[1], **Dr Rosemary Thresher**
[1]*LIC, Hamilton, New Zealand*

Epidemiology and Economics - Poster Pitches, June 16, 2022, 12:15 PM - 12:30 PM

It is estimated that 60% of New Zealand dairy herds have Johne's disease (JD). Despite this, at present no formal JD control program exists. Practices to mitigate transmission risks or identify positive animals are the responsibility of the farmer and their veterinarian.
The majority of New Zealand herds carry out herd testing at least once per season, in a process where a milk sample is collected from each individual milking animal. These samples are analysed for a number of production traits, and offer the farmer insights into individual animal productivity and health.
Livestock Improvement Corporation (LIC) has offered an antibody ELISA to identify Mycobacterium avium subspecies paratuberculosis (MAP) antibody-positive animals through herd test milk since 2013. Since initial release, the herd test MAP ELISA has seen significant growth. Nearly 800,000 herd test milk samples were analysed on this diagnostic platform in the 2020-21 season, which equates to 16.3% of all dairy cows recorded in NZ.
A higher proportion of South Island herds were JD tested, with nearly 1 in 4 utilising the herd test milk ELISA in the 2020-21 season, covering more than a quarter of the total animals (28.2%). North Island testing covered 6.2% of herds, and 7.6% of animals.
Testing for JD across several seasons appears to be having an impact on overall antibody-positive prevalence within herds. In 2020, herds who had routinely tested for a minimum of 3 seasons tended to have lower prevalence than those who were testing for the first time.
Further work utilising this dataset for insights into JD in New Zealand is underway.

322

Invited Speaker for Public Health and Map in the Environment

Professor Marcel Behr

Public Health and Map in the Environment - Invited Speaker, Main Auditorium, Printworks, June 16, 2022, 1:30 PM - 2:15 PM

Five learning points

1. The Tortoise and the Snail: Is Science on MAP and Crohn's Moving Slowly or Very Slowly? Studies looking for an association between Mycobacterium avium paratuberculosis (MAP) and Crohn's disease should use validated laboratory methodologies.

2. Labs using these methodologies to look for MAP in human samples should subject themselves to rigorous quality control, to ensure that their findings are robust.

3. Clinical trials that monitor response to antibiotics do not formally address the MAP hypothesis; rather, they ask whether a defined treatment makes patients better.

4. Clinical trials that have embedded validated laboratory techniques can examine whether there is an association between MAP status and treatment response.

5. Clinical trials that screen by MAP status and then randomize patients could provide evidence of a treatment benefit in MAP positive patients; to my knowledge, such trials have not been done.

189

Heat resistance of Mycobacterium avium subsp. paratuberculosis: inactivation kinetics during the production process of Mozzarella cheese

Dr Filippo Barsi[1], Dr Elena Cosciani-Cunico[2,4], Dr Simone Russo[1], Dr Elena Dalzini[2,4], Dr Anita Filippi[1], Dr Giuliana Cammi[1], Dr Norma Arrigoni[1], Dr Paolo Daminelli[3], Dr Nadia Losio[2,4], Dr Chiara Garbarino[1], Dr Matteo Ricchi[1]
[1]IZSLER-Sede di Piacenza - National Reference Laboratory for Paratuberculosis, Via Strada Faggiola, 1 - Podenano (PC), Italy, [2]IZSLER-National Reference Centre for Emerging Risks in Food Safety, Via Celoria, 12 - Milan, Italy, [3]IZSLER-Reparto Produzione Primaria Centro di IZSLER - National reference center for the quality of bovine milk, Via Bianchi, 5 - Brescia, Italy, [4]IZSLER-Food Control Division of the Food Safety Department, , Via Bianchi, 5 - Brescia, Italy

Public Health and Map in the Environment - PM Session (1), Main Auditorium, Printworks, June 16, 2022, 2:15 PM - 3:00 PM

The hygienic guarantees requested by some countries relative to MAP presence on imported dairy products should be supported by studies demonstrating the efficacy of MAP inactivation during the technological processes. The Mozzarella cheese is one of the most exported Italian cheeses and the definition of inactivation kinetics enables the theoretical evaluation of thermal processes for a better definition of MAP heat resistance. Thus, the aim of this study was to investigate MAP inactivation kinetics in curd during the production of this cheese.
A challenge test has been carried out spiking pasteurized cow milk with ATCC 19698 strain. To simulate the heating/stretching of the curd, the critical step for pathogens inactivation during the Mozzarella production process, the mature curd was under-vacuum packed and treated, separately, in hot water at six different temperatures (ranging from 60°C to 75°C) for a maximum of 12 min. The MAP survival was analysed by plate count method in HEY medium supplemented with Mycobactin and the obtained survival curves were fitted using the Geeraerd-tail model in the "add-in GInaFit" suite in order to estimate the Kmax (specific inactivation rate) for each temperature. The kinetic thermal inactivation parameters, such as the D-value (time to inactivate 90% of the cells at a given temperature) and the Z-value (increase in temperature inducing a 10-fold reduction in D-value) were calculated by these data.
The Kmax for the MAP ranged from 0.63 (min-1) at 60°C to 15.10 (min-1) at 75°C in stretched curd, showing D-values from 3.6 min to 0.15 min respectively at 60°C and 75°C with a Z-value of 10.3°C, confirming the high heat resistance of MAP.
Our data show the knowledge of kinetic parameters regulating MAP survival during curd thermal treatments have a deep impact in the safety of mozzarella cheeses and can be useful for all stakeholders.

242

Effluent as a herd-level screening tool for the estimation of Johne's disease prevalence within New Zealand milking herds

Miss Rebecca White, Dr Shona Pryor[1], **Dr Rosemary Thresher**
[1]*Livestock Improvement Corporation, Hamilton, New Zealand*

Public Health and Map in the Environment - PM Session (1), Main Auditorium, Printworks, June 16, 2022, 2:15 PM - 3:00 PM

In the pasture based dairy systems of New Zealand, faeces is collected from yards into bulk effluent storage areas. Sampling of effluent from these sites is a convenient way to screen for Mycobacterium avium subspecies paratuberculosis (MAP) as a herd level indicator of Johne's disease. The ability of a derived effluent MAP score to estimate prevalence of MAP antibody-positive animals in the milking herd was trialled.
Effluent samples were collected from 2-4 locations/farm in 49 farms across New Zealand. 6 farms were analysed twice, giving a total of 55 sampling events. DNA was extracted and analysed for MAP by real time PCR. An effluent MAP score was calculated depending on the signal strength of all the effluent samples, then a category assigned of either 'not detected', 'low', 'moderate to high', or 'very high' based on bacterial burden.
Individual milk samples were collected from the milking herd at a similar time to the effluent sampling and analysed using a MAP Antibody ELISA. The effluent MAP score was then compared with the percentage prevalence of antibody-positive animals.
Prevalence of antibody-positive animals was significantly higher for farms with a "very high" compared with "moderate to high" MAP effluent category (P=0.006). Similarly, prevalence was significantly higher when comparing "moderate to high" with "not detected" (p=0.001). A 'low' MAP result was recorded in 3 herds, excluding this category from statistical evaluation.
Repeat sampling of low prevalence herds (<0.5% antibody-positive) demonstrated switching between MAP being detected ('low' or above) and not detected. In at least one instance, a faecal super shedder confounded the predictive ability of the MAP score, as a higher than expected bacterial load was identified in the effluent when compared to antibody-positive prevalence.
Work is ongoing to further understand test accuracy, particularly in low/no prevalence herds, and in the presence of super shedders.

198

Presence of Mycobacterium avium spp. paratuberculosis in silage collected on commercial cattle farms

Dr. Susanne Eisenberg[1], Dr. Amely Campe[2], Dr. Karsten Donat[3,4]
[1]Animal Disease Fund of Lower Saxony, Hanover, Germany, [2]University of Veterinary Medicine Hanover, Hanover, Germany, [3]Thuringian Animal Disease Fund, Jena, Germany, [4]Justus-Liebig-Universität Gießen, Gießen, Germany

Public Health and Map in the Environment - PM Session (1), Main Auditorium, Printworks, June 16, 2022, 2:15 PM - 3:00 PM

Mycobacterium avium spp. paratuberculosis (MAP) belongs to the group of acid-fast bacteria which can be identified using a Ziehl-Neelsen staining. Therefore, silage has been named as a possible risk in MAP transmission. The objective was to study the presence of MAP in silage samples and the correlation with MAP status of the herd. MAP-positivity was defined as “at least one MAP-positive environmental sample”. It was hypothesized that only a small number of silage samples are MAP positive and that herds with a positive MAP status were more likely to have MAP-positive silage samples. Between July 2017 and January 2019, silage and environmental samples were collected at 107 cattle farms located in Lower Saxony and Schleswig-Holstein. Silage samples were analyzed for the presence of MAP DNA by IS900 realtime PCR, whereas slurry, boot swab and environmental fecal samples were analyzed by fecal culture and IS900 realtime PCR. Presence of MAP in silage as well as in environmental samples was recorded as a binary outcome (MAP detected yes or no). Proportions were compared using Fisher exact statistics. In total 243 silage samples and 204 environmental samples were analyzed. 5 out of 243 silage samples belonging to 5 farms were found PCR positive. 25 out of 205 environmental samples belonging to 15 farms were found PCR positive and 19 were culture positive. Of 107 herds, 15 were classified as MAP positive. However, only 1 out of 5 MAP PCR positive silage samples originated from MAP-positive herd. A correlation between herd-MAP status and positive silage samples was not identified. In conclusion, only a small number of silage samples were detected as MAP positive and therefore, feeding silage poses only a small risk in MAP transmission compared to a contaminated environment.

266

Detection of Mycobacterium avium subspecies paratuberculosis in environmental samples from a dairy goat farm in Germany

Miss Chris Pickrodt[1], Dr. Karsten Donat[2], Dr. Heike Köhler[1]
[1]Institute of Molecular Pathogenesis, Friedrich-Loeffler-Institut (Federal Research Institute for Animal Health), Jena, Germany, [2]Thuringian Animal Diseases Fund, Animal Health Service, Jena, Germany

Public Health and Map in the Environment - PM Session (2), Main Auditorium, Printworks, June 16, 2022, 3:30 PM - 4:45 PM

This study assessed the occurrence of Mycobacterium avium subspecies paratuberculosis (MAP) in the barn environment of a dairy goat herd with paratuberculosis history two years after introducing vaccination with Gudair.
Bedding, dust, water and feed samples were collected and analysed for viable MAP and MAP DNA. Sample collection took place in May at the end of lambing season just before turning out to graze as well as in December after cleaning the barn. DNA extraction and qPCR were performed with commercial kits according to manufacturer instructions. Additionally, all samples were cultivated on Herrold's Egg Yolk Medium over 24 weeks.
MAP was cultivated out of 8/36 bedding and 1/14 dust samples. 3/8 water, 16/36 bedding and 10/14 dust samples showed qPCR positive results. There was no evidence of MAP in the feed samples (0/2). Intense animal traffic areas as the entrance to the barn (Ct ≈ 32), the rotary milking parlour and the waiting pen (Ct ≈ 33) showed a high bacterial load. MAP DNA was detected in water and dust samples from the kid-rearing area probably due to airborne distribution of the pathogen. More positive results were obtained from the lambing pen of multiparous goats (parity >1) compared to the pen of primiparous goats. Fewer positive results in the lambing area in December point to an effective reduction of MAP by cleaning and disinfection.
Despite the herd had been vaccinated for two years MAP was found in the farm environment of the goat herd demonstrating that vaccination as a solitary measure of disease control does not lead to an effective reduction of environmental contamination. The identification of areas with high risk of MAP contamination in small ruminant husbandries suggests the implementation of targeted hygiene measures to reduce the pathogen's prevalence and to control the disease.

205

Environmental free-living amoebae interaction with Mycobacterium avium subsp. paratuberculosis

Mr Amélie Jessu[1], Thierry Cochard[2], Ascel Samba-Louaka[1], Jean-Louis Moyen[3], Franck Biet[2], Yann Héchard[1]
[1]Laboratoire Écologie et Biologie des Interactions (EBI), Team Microbiologie de l'Eau, Université de Poitiers, UMR CNRS 7267, Poitiers, France, [2]INRAE, ISP, Université de Tours, Nouzilly, France, [3]Laboratoire Départemental d'Analyse et de Recherche de la Dordogne, Coulounieix-Chamiers, France

Public Health and Map in the Environment - PM Session (2), Main Auditorium, Printworks, June 16, 2022, 3:30 PM - 4:45 PM

Free-living amoebae (FLA) are ubiquitous in environment and feed on bacteria by phagocytosis. Depending on the mycobacteria species, they can be digested, survive or grow within FLA. Studies have shown that Mycobacterium avium subsp. paratuberculosis (Map) can grow in FLA in vitro but too few have studied this interaction in an environmental context. Understanding its survival and its persistence in environment is important to better understand its transmission cycle. Our hypothesis is that FLA can be an environmental reservoir and vector of Map.

To test this hypothesis, water and soil samples from infected farm were used to culture FLA, which were lysed to extract total DNA and intracellular bacteria. Total DNA were used to detect Map by a nested qPCR. Lysates were used to culture Map. For all qPCR-positive samples, FLA were identified and Map genotyped. Environmental FLA were finally used to characterise the interaction with Map by infection assays.

For all samples, we isolated FLA. The nested qPCR method developed showed a higher sensitivity than the classical qPCR, allowing us to detect trace of Map DNA in the total DNA. Several samples were positive for qPCR-positive for Map. This underlined that at least Map DNA could be found in association with FLA. The identification of Map and FLA strains are in progress and will be presented. The permissiveness of FLA to Map is investigated by infection assays using a FLA isolated from cattle environment, Rosculus like, and with newly isolated FLA.

Our work led to isolate many environmental FLA from infected farms, improve the sensitivity of Map detection by nested qPCR, confirm the presence of Map DNA in interaction with environmental FLA and follow the permissiveness of these FLA to Map infection. These results highlight the possible role of FLA as reservoirs of Map in the environment.

255

Cumulative sensitivity and specificity of repeated environmental sampling and milk-pool testing as surveillance methods at herd level in large low-prevalence dairy herds

Dr. med.vet. habil. Karsten Donat[1,2], Annika Wichert[1], Dr. Esra Einax[1], Elisa Kasbohm[3]

[1]Thuringian Animal Disease Fund, Victor-Goerttler-Strasse 4, Jena, Germany, [2]Justus-Liebig-University, Clinic for Obstetrics, Gynecology and Andrology with Veterinary Ambulance , Gießen, Germany, [3]University of Greifswald, Institute of Mathematics and Computer Science, Greifswald, Germany

Public Health and Map in the Environment - PM Session (2), Main Auditorium, Printworks, June 16, 2022, 3:30 PM - 4:45 PM

This study aimed at applying repeated environmental sampling and milk-pool testing as surveillance methods at herd level within a Johnes' control programme and to estimate cumulative sensitivity and specificity of repeated testing. The study was conducted in 36 Thuringian dairy herds with known within-herd prevalence (WHP) of MAP-shedders as estimated from the results of annual testing of the herds' cows by individual faecal culture or qPCR. In each study herd, twelve boot swab samples, environmental faecal samples and liquid manure samples, respectively, were taken at monthly intervals over the period of one year. Each environmental sample was analysed separately by both bacterial culture and faecal qPCR. Besides, all environmental samples (except the boot swab sample) taken at the same on-farm visit were pooled and analysed by bacterial culture and qPCR. In addition, individual test day milk samples of all lactating cows were pooled in groups of 25 and 50 and tested by a commercial ELISA for MAP-specific antibodies. In MAP-free study herds, testing for MAP using environmental samples did not yield in any false MAP-positive result, indicating a specificity of 100%. In contrast, MAP-positive milk pools originating from MAP-free herds resulted in a reduced cumulative specificity. Herds with a WHP ≥ 2% were detectable with a sensitivity ≥ 95% using pools of repeatedly taken environmental samples. In herds with a WHP < 2%, none of our applied methods was able to achieve the aim to detect MAP-positive herds with a sensitivity ≥ 95%. In conclusion, repeated sampling of environmental samples and pooled milk samples is an effective method to detect large MAP-positive dairy herds with a WHP ≥ 2% but not < 2% with sufficient confidence. Therefore, in the context of certifying herds as MAP-free, a combination of repeatedly collected different samples and trade control is necessary and recommended.

270

Therapeutic management of Mycobacterium avium subspecies paratuberculosis infection with complete resolution of disease in patients with Inflammatory Bowel Syndrome in India

Dr Saurabh Gupta[1,4], Dr Kundan Kumar Chaubey[1,4], Dr Prabhat Agarwal[3], Dr Ashish Gautam[3], Dr Manju Singh[2], Dr P K Singh[1], Dr Shoor Vir Singh[1,4]

[1]Department of Biotechnology, GLA University, Mathura, Mathura, India, [2]Adesh Institute of Medical Sciences & Research, Adesh University, Bathinda, Bathinda, India, [3]Department of Medicine, S.N. Medical College, Agra, Agra, India, [4]ICAR-Central Institute for Research on Goats, Makhdoom, Mathura, India

Public Health and Map in the Environment - PM Session (2), Main Auditorium, Printworks, June 16, 2022, 3:30 PM - 4:45 PM

Background: Mycobacterium avium subspecies paratuberculosis (MAP) infection, the cause of Johne's disease (Paratuberculosis) is endemic in the domestic livestock population. MAP is excreted through milk therefore continuously entering human food chain from infected cows, buffaloes and goats. MAP has been associated in humans beings suffering with autoimmune diseases such as Crohn's disease, Type 1 diabetes, Hashimoto's thyroiditis, Rheumatoid arthritis, etc. Patients had a history of frequent bowel movements, loose stool with mucus and loss in body weight and diagnosed as a cases of inflammatory bowel disease (confirmed by colonoscopy and radiographic imaging). Patients did not respond to routine and standard therapy.

Methods: Screening for the presence of MAP was done using stool culture, microscopy, IS900 PCR and Indigenous ELISA test. Patients were heavily shedding typical acid fast MAP bacilli (+3 to +4) by microscopy and had low titre against MAP-antibodies by 'indigenous ELISA kit'. MAP stool culture was positive for all the IBD patients after 6 months incubation. Anti-MAP therapy was started under the supervision of physician. Treatment regimen included isoniazid, rifampicin/ rifaximin, Rifabutin ,ethambutol once a day; and Clathriomycin/Levofloxacin twice a day along with anti-inflammatory drug mesalamine.

Results: Following treatment, patients progressively improved with reduction in bowel movement frequency, gained in body weights and improvements in appetite. After one year of treatment, stool microscopy and culture were negative for MAP. Patients are being monitored for their health and probable relapse.

Conclusions: Patients have experienced complete resolution of IBD using a combination of anti-tuberculosis antibiotics. These case reports are one more example of the linkage of demonstrable MAP infection treated with anti-tuberculosis therapy to the presence and then absence of disease in the humans.

169

Combining optimized sample extraction with sensitive detection using bactotype® MAP real-time PCR

Nikita Tandon[1], Dr Christine Gaunitz[1], Dr Marco Labitzke[1], Dr Claudia Engemann[1], Oliver Sasse[1], Leslie Moussi[1], Dr Fredrik Ullman[1]
[1]*INDICAL BIOSCIENCE GmbH, Leipzig, Germany*

Introduction: The purpose of this study is to show increased sensitivity of Mycobacterium avium subsp. paratuberculosis (MAP) detection by combining sensitive, specific amplification with optimized protocols for the extraction of MAP DNA from fecal samples.
Direct fecal PCR is becoming widely used since it provides test results within hours. The challenges of MAP DNA extraction from fecal samples include MAP cluster variation in the sample, thick mycobacterial cell walls, and PCR inhibitors.

Materials and Methods: Different fecal sample pretreatments, combining chemical lysis and mechanical disruption were tested. MAP DNA extraction was performed using the magnetic bead based IndiMag Pathogen Kit.
For the detection of MAP DNA, bactotype MAP PCR Kit was used. The kit is a duplex real-time PCR including a ready-to-use master mix and a heterologous extraction and amplification control. Analytical sensitivity of the bactotype MAP PCR Kit was tested using titration series of in vitro DNA [106–100 copies/well]. The performance of the MAP workflow was evaluated based on samples of USDA Proficiency Panels and samples from different European ring trial panels and field samples.

Results: Clear differences were shown between different homogenization matrices, lysis buffers and protocols. In combination with F-MAP pretreatment, bactotype MAP PCR Kit detected all ring trial and proficiency test samples correctly.
Samples, positive in another commercial MAP PCR workflow, were detected positive with
lower CT values, indicating better sensitivity of bactotype MAP PCR Kit. Up to 5 MAP DNA copies per sample were detected with high efficiency including weak positive samples from low shedders.

Conclusion: For the detection of MAP in bovine fecal samples, a good combination of optimized sample preparation and a sensitive real-time PCR is key. The bactotype MAP PCR Kit in combination with INDICAL's extraction kits allows reliable and fast MAP diagnosis from ruminant fecal samples as part of control programs.

180

Evaluation of fecal culture, fecal RT-PCR, three commercially available serum ELISA's, complement fixation and AGID assays as diagnostic tools to detect Mycobacterium avium ssp. paratuberculosis in dairy sheep

Jennifer Cooper[1], Dr. Jessica Scillieri Smith[2], Dr Nicole Aulik[1]

[1]*Wisconsin Veterinary Diagnostic Laboratory-uw Madison, Madison, United States,* [2]*New York State Department of Agriculture and Markets, Albany, United States*

Paratuberculosis is a chronic infection in small ruminants caused by Mycobacterium avium subspecies paratuberculosis (MAP). MAP has been shown to be major production-limiting disease in sheep and goats. This study's objective was to evaluate three commercially available serum ELISA's, AGID and complement fixation for the antibody detection of MAP in sheep as possible surveillance and diagnostic assays. Fecal samples collected at the same time as blood samples, were also tested to evaluate four RT-PCR targets (IS900, F57, ISMAP02, hspX) and culture followed by PCR as possible diagnostic tools for MAP detection in sheep. Paired fecal and blood samples were collected between March and November 2019 from 21 different farms. Results from the paired samples and subsequent re-sampling of select animals showed that utilization of a single testing strategy for the surveillance or diagnostic evaluation of a flock is not recommended due to the intermittent shedding of target in feces and the un-reliability of serology assays to correctly identify positive animals.

181

Detection of a Mycobacterium avium spp avium in a pudu and muntjac in the United States

Jennifer Cooper[1], Dr Adel Talaat[2], Dr Adrian Mutlow, Dr Kenneth Conley, Dr Nicole Aulik[1]

[1]Wisconsin Veterinary Diagnostic Laboratory-uw Madison, Madison, United States, [2]UW School of Veterinary Medicine, Madison, Unites States

Recently, a pudu and muntjac, were identified as having a disease similar to Johne's disease including diarrhea and wasting. A fecal sample from the pudu tested positive by a commercially available, ISMAP02 RT-PCR for Mycobacterium avium subspecies paratuberculosis (MAP). The positive result and continued wasting prompted euthanization and submission of tissues, serum and feces. The samples were analyzed using different MAP-specific PCR targets, which yielded no positive results. However, three MAP antibody ELISAs demonstrated a positive result for MAP. Culture was performed on the samples and revealed acid-fast positive rods, which were confirmed by histology. PCR analysis of the culture did not result in positive PCR results for three targets to MAP. In juxtaposition, the muntjac died and tissues and feces were submitted for PCR and culture. PCR did not yield positive results, but similar to the pudu, the culture showed acid fast rods, which were PCR negative for MAP targets. To identify the acid-fast rods, 16S and whole genome sequencing (WGS) was employed and showed these strains to be most closely related to Mycobacterium avium subspecies avium (MAA). This data demonstrates that MAA can cause disease in a pudu and muntjac, but interestingly the pudu isolate cross-reacted with a common PCR target and gave false-positive results. Preliminary analysis shows that there may be a genetic recombination occurring in these exotic species as both MAA had significant single nucleotide differences (SNPs) from the reference MAA strain and the isolates also had significant differences between each other. This data also indicates that it is imperative that surveillance for MAP in exotic species must include a multiplex PCR with more than one target. If a disease similar to MAP is suspected in an exotic species, culture and WGS may be necessary to identify the causative agent.

192

Detection of antibodies against Mycobacterium avium subspecies paratuberculosis in sera and bulk tank milk samples from French goat herds.

Dr Virginie Thibault-Poisson[1], Ms Mélanie Thirion[1], Ms Lucie Pineau[1], Ms Isabelle Bremaud[1], Mr Marc Tabouret[1]
[1]French Agency for Food, Environmental and Occupational Health and Safety (Anses), Ploufragan-Plouzané-Niort Laboratory, Unit Pathology and Welfare of Ruminants, Niort, France, [2]Animal health protection farmer's organization (GDS France), Paris, France

Introduction: In goats, paratuberculosis results in progressive weight loss, production drops and diarrhea. In the absence of treatment, the control of paratuberculosis is essentially based on health measures (reform of infected animals and hygiene) associated or not with vaccination depending on the intra-herd prevalence of the infection.
The purpose of the present study was to assess the seroprevalence of paratuberculosis in caprine herds and to evaluate its correlation with Mycobacterium avium subsp. paratuberculosis (Map) antibody level in bulk tank milk.

Materials and Methods: A set of 8267 animals from 81 herds was selected in the Poitou-Charente region, France, between June 2012 and January 2013. Map antibody detection was realized on individual sera from randomly selected goats and on bulk tank milk using different commercial ELISA kits.

Results, discussion and conclusion: The estimated intra-herd seroprevalences were comprised between 0% and 21,9% with a median at 6,5%. Nineteen (23,5%) herds had a prevalence of 0% and sixteen (19,8%) had a prevalence above 15%. Results obtained on bulk tank milk from the same herds showed a good correlation with the seroprevalence whatever the ELISA kit used. Optimization of ELISA cut-off levels was possible to improve the sensitivity of the detection of herds with a low number of positive goats.
Map antibody detection performed on bulk tank milk could be an interesting and inexpensive tool for the screening of herds with various prevalence of infected goats.

193

Interactions Between Mycobacterium avium subsp. paratuberculosis and Staphylococcus aureus and Streptococcus agalactiae, in Epithrlial Cells of the bovine Mammary Gland - Mac-T

MSc Junnia Luísa Pena[1], DSc David Germano Gonçalves Schwarz[2], DSc Arthur William Brasil[3], DSc Leandro Licursi de Oliveira[1], DSc Abelardo Silva Júnior[1], **PhD Maria Aparecida Scatamburlo Moreira[1]**

[1]Universidade Federal de Viçosa, Viçosa, Brazil, [2]Universidade Federal do Piauí, Bom Jesus, Brazil, [3]Universidade Federal de Rondônia, Rolim de Moura, Brazil

The mammary gland contributes as the main route of MAP transmission to newborn and human animals. Although not directly related to mastitis, MAP may be influencing the microorganisms that cause this disease. The objective of this research was to evaluate the influence of MAP in coculture with Staphylococcus aureus and Streptococcus agalactiae on MAC-T cells. Bacterial internalization, colony counting, molecular assays and MAC-Ts viability evaluation were performed. In the internalization test between MAP and S. aureus, there was a 30 minute time difference between MAC + S. aureus and MAC-T + MAP + S. aureus treatments. MAC-T + S. aureus treatment showed significant differences between 10 and 120 minutes ($p = 0.001$) and between 30 and 120 minutes ($p = 0.048$). MAC-T + MAP + S. aureus treatment showed a difference between 10 minutes and 30 minutes ($p = 0.042$) and between 10 and 120 minutes ($p <0.0001$). In coinfection performed with S. agalactiae, the tests showed no significant difference between treatments. MAC-T + S. agalactiae treatment showed significant differences between 10 and 120 minutes ($p <0.0001$) and between 30 minutes and 120 minutes ($p = 0.030$). MAC-T + MAP + S. agalactiae treatment showed a difference between 10 and 120 minutes ($p <0.0001$) and between 30 and 120 minutes ($p = 0.029$). The cellular viability of MAC-T was verified by the MTT test, concluding that the internalization process was not influenced by the death of MAC-Ts, but by the presence of MAP, S. aureus and S. agalactiae. These results show interaction between MAP and S. aureus and S. agalactiae, favoring the internalization of S. aureus and not influencing the invasion of S. agalactiae during internalization tests, which may contribute to the worsening of mastitis and permanence of these agents inside. of mammary epithelial cells, compromising the treatment of mastitis.

202

Results of fifteen years Milk Quality Assurance for Paratuberculosis in Dutch dairy herds

Dr. Maarten F. Weber[1,2], Dr. Marian Aalberts[1], Dr. Thomas Dijkstra[1], Prof. Ynte H. Schukken[1,2,3]

[1]Royal GD, Deventer, the Netherlands, [2]Dept. of Population Health Sciences, Faculty of Veterinary Medicine, Utrecht University, Utrecht, the Netherlands, [3]Quantitative Veterinary Epidemiology, Dept. of Animal Sciences, Wageningen University, Wageningen, the Netherlands

In 2006, a milk quality assurance programme (MQAP) for paratuberculosis in Dutch dairy herds was initiated to reduce the concentration of Mycobacterium avium subsp. paratuberculosis (Map) in bulk milk delivered to the milk processors. The aim of this study was to evaluate results achieved in a 15-year period (2006-2020) in 718 dairy herds that entered the programme in 2006 and 2007.

Herds participating in the MQAP are assigned a herd status based on the results of herd examinations. Each herd examination consists of testing either individual milk samples of all lactating cattle or serum samples of all cattle ≥ 3 years of age by ELISA. Farmers are entitled to confirm positive ELISA results by individual faecal PCR assay or culture.

An initial assessment of participating herds consists of a single herd examination. Test-negative herds enter a surveillance procedure and are assigned status A. The surveillance of status A herds consists of biennial herd examinations. Test-positive herds enter a control procedure and are assigned status B (if all test-positive cattle have been removed from the herd) or status C (otherwise). If an annual herd examination in a herd with status B yields negative results only, the herd progresses to status A.

During the 15-year period the proportion of participating herds with the preferred herd status (status A) increased from 45% to 80%. Moreover, a decreasing apparent prevalence (across all age groups as well as in heifers at their first test) and an increasing age at onset of test-positivity were observed (age before which 10% of cattle became test-positive increased from 6 to 8 years). These observations are indicative of a reduced transmission of Map after long-lasting participation in the MQAP. The results of this study indicate that the MQAP positively contributes to the control of Map in the Dutch dairy herds.

204

Performance of three commercial ELISA tests for Mycobacterium avium subsp. paratuberculosis (MAP) antibodies in blood and milk samples

Dr. Ingrid Lorenz[1]

[1]*Bavarian Animal Health Service, Poing, Germany*

The purpose of this study was to investigate and compare the performance of three different commercial ELISAs for the detection of antibodies against MAP in blood and milk from cows of known positive farms in different stages of control efforts.

Blood, milk and fecal samples were collected from cows over two years of age from eight Bavarian dairy farms with confirmed infection with MAP. A cow was defined as positive if MAP could be detected on fecal culture. Three commercial ELISAs (ID Screen Paratuberculosis Indirect, ID.vet (1); IDEXX Paratuberculosis Verification, IDEXX Laboratories (2); cattletype MAP Ab, Indical Bioscience (3)) were used to test for MAP antibodies in serum and milk samples. IBM SPSS Statistics 24 was used for statistical analysis. ROC analysis was used to compare tests and test combinations.

804 serum samples (44 from MAP positive cows) were analyzed. The area under the curve (95% confidence intervall) was 0,856 (0,791 - 0,921), 0,862 (0,791 - 0,933) and 0,825 (0,791 - 0,933) for tests 1, 2 and 3, respectively. For the combination of tests 1 and 2 values were 0,882 (0,823 - 0,941), other combinations did not improve performance. 1688 milk samples (93 from MAP positive cows) were analyzed. The area under the curve (95% confidence intervall) was 0,863 (0,813 - 0,913), 0,863 (0,814 - 0,911) and 0,822 (0,772 - 0,872) for tests 1, 2 and 3, respectively. In milk samples combination of tests did not improve performance.

Reliability of the investigated ELISAs was acceptable to good and did not differ between serum and milk sample. Test 3 performed slightly worse than test 1 and 2. Best results were achieved with a combination of test 1 and 2 in serum samples.

209

Is copper ions treatment enough to stop Mycobacterium avium subsp. paratuberculosis (MAP) infection progression in newborn calves?

Dr Miguel Salgado[1], Dr Pamela Steuer[1], Mr Carlos Tejeda[1], Mr Manuel Moroni[1], Dr. Cristobal Verdugo[1]

[1]*Instituto de Medicina Preventiva Veterinaria, Facultad De Ciencias Veterinarias, Universidad Austral De Chile, Valdivia, Chile*

Introduction: Milk is a very important transmission route of MAP. This pathogen can survive pasteurization, being necessary another alternative to control MAP infection through this pathway. Several studies have shown the antimicrobial properties of copper. The present study reports the application of a novel milk decontamination treatment based on copper ions to evaluate the course of MAP infection on calves fed with copper-treated milk.

Material and Methods: A one-year field longitudinal study was carried out. Newborn calves were assigned to one of four experimental groups, where one received copper treated milk, another contaminated milk (104 MAP/mL), and the other two were negative controls. Milk samples were taken to estimate copper effectiveness. Additionally, calves' fecal samples along the study period were taken, in order to monitor infection progression. Finally, calves were euthanized, and tissue samples were taken for anatomo and histopathological analysis.

Results: MAP load in treated milk significantly decreased. Fecal elimination of MAP was observed in all study groups, but the group of calves naturally exposed to MAP infection began to shed earlier than the rest. Only calves fed with copper treated milk showed evidence at the histopathological level consistent with MAP infection.

Conclusion: Treatment with copper ions significantly reduced MAP loads in milk. However, only the copper treated milk calves group showed confirmatory signs of MAP infection progression. The latter offers more questions than answers, and maybe the presence of a more tolerant and virulent MAP strain could be the final answer to this situation.

214

Twenty-five topics to be considered in the design of paratuberculosis control programmes for dairy herds

Dr. Maarten F. Weber[1,2], Dr. Thomas Dijkstra[1], Dr. Marian Aalberts[1], Prof. Dr. Ynte H. Schukken[1,2,3]
[1]Royal GD, Deventer, the Netherlands, [2]Dept. of Population Health Sciences, Faculty of Veterinary Medicine, Utrecht University, , the Netherlands, [3]Quantitative Veterinary Epidemiology, Dept. of Animal Sciences, Wageningen University, , the Netherlands

In 2006, a milk quality assurance programme (MQAP) for paratuberculosis in Dutch dairy herds was initiated to control Mycobacterium avium subsp. paratuberculosis (Map) and reduce the concentration of Map in milk delivered to milk processors.

Although the results achieved to date are promising, an internal review of the MQAP was performed to identify potential areas of improvement to increase its efficacy. In this review, progress made in controlling Map was summarized whilst the design of the MQAP was compared with programmes in other European countries. The literature on control programmes, epidemiology, diagnostics, zoonotic aspects and on-farm impact of Map that was published since the initiation of the MQAP in 2006 was reviewed for knowledge and developments that could potentially warrant a revision of the MQAP.

In all, 25 potential areas of improvement were identified and grouped into nine domains: (i) aims and periodical evaluation of the programme, (ii) knowledge, attitude and behaviour of farmers and veterinarians, (iii) surveillance of herds with the preferred herd status, (iv) herd examinations in test-positive herds, (v) progress of previously test-positive herds to the preferred herd status, (vi) preventive management measures, (vii) animal movements between herds, (viii) certification and surveillance of designated young stock herds, and (ix) breeding for resistance against Map.

The review was presented to decision makers of the Dutch dairy industry, who concluded that the identified topics in the design of the MQAP were valuable for future consideration. Moreover, the 25 identified topics may serve as a valuable checklist of items that might be considered in the design of paratuberculosis control programmes in other countries.

220

The biosecurity and bio-containment risks associated with MAP infection and prevalence in UK dairy herds

Mr Richard Sibley[1], Mr Peter Orpin[1]
[1]*West Ridge Vets, Witheridge, United Kingdom*

The risks of disease entry and spread in over 2500 UK dairy herds using a risk assessment tool to measure, monitor and manage risks were studied to determine risks associated with infection and prevalence of MAP within the herd.
A web based risk assessment tool was used by trained veterinary surgeons to assess, measure and monitor biosecurity and biocontainment risks on over 2500 UK dairy herds that engaged in some form of Johnes management. The tool uses standard assessments and an algorithm to quantify risks and provide a summary and priority to identify and manage risks as part of the control plan.
50% of herds were designated high risk of entry of Johnes disease in to the herd, mostly due to purchasing cattle of unknown disease status and allowing cattle to drink from water courses that had passed through other livestock farms. Only 20% of dairy herds were designated as low biosecurity risk for the entry of Johnes disease.
65% of herds were designated as high risk of spread, mostly due to the use of multiple calving areas and poor perinatal hygiene, and the use of pooled colostrum taken from high risk cows. Only 8% of dairy herds were designated as low risk of spread.
Of 2462 herds that were designated as having high risks of spread of MAP within their herds, 52% had high risks of entry of the disease, making these herds very high risk of a high prevalence of Johnes disease. 85% of these high risk herds were infected as determined by milk elisa testing. The study demonstrates that herd infection is associated with biosecurity risks and herd prevalence is associated with biocontainment risks.

224

Prediction of clinical paratuberculosis with RT-qPCR

Dr Barbara Vass-Bognár[1], Dr Walter Baumgartner[2], Kinga Fornyos[3], Dr Viktor Jurkovich[1]
[1]University of Veterinary Medicine, Budapest, Hungary, [2]University of Veterinary Medicine, Vienna, Austria, [3]M.A.H. Food-Control Ltd, , Hungary

This work aimed to measure the detectability of paratuberculosis (PTB) during the lactation period. Our goal is to find time intervals when one can predict the disease with better efficacy. The culling of animals before the onset of clinical symptoms is desirable from both an economic and a veterinary point of view.
We carried out our investigations on 4 Hungarian dairy farms with an average of 597 milking cows per farm. The apparent prevalence of PTB infection based on serum ELISA examinations was 7.6 (± 3.9) %. Faeces samples were taken at 5, 10-14, 40-60, 90-120 and 200 days after calving from 38 animals in total. Real-time qPCR using Macherey-Nagel NucleoSpin Tissue and Adiavet Rt-PCR kit was performed in the faeces samples. Nine cows were culled because of clinical symptoms of PTB. All cows with clinical signs had a positive RT-qPCR result with lower Ct (cycling threshold) than 24. The animals started to show clinical signs 14-20 days after calving, and they were sent to the slaughterhouse on average 60 days after calving. The average Ct value in the faeces of the culled animals was 20.2.
Based on our results, faecal RT-qPCR after calving is a possible predictor of disease outcome in later stages of lactation and can support the culling protocol in the PTB eradication programs.
Acknowledgements: Viktor Jurkovich was supported by the Ministry of Innovation and Technology and the National Research, Development, and Innovation Fund (grant No.: MEC_21_R_140844).

226

Evaluation of single nucleotide polymorphisms (SNPs) associated to genetic resistance to paratuberculosis in marchigiana beef cattle

Doctor Piera Mazzone[1], Doctor Linda Petrucci[1], Doctor Martina Torricelli[1], Doctor Sara Corneli[1], Doctor Antonella Di Paolo[1], Doctor Carla Sebastiani[1], Doctor Marcella Ciullo[1], Doctor Martina Sebastianelli[1], Doctor Silva Costarelli[1], Doctor Eleonora Scoccia[1], Doctor Fiorella Sbarra[2], Doctor Federica Gabbianelli[3], Professor Alessio Valentini[3], Doctor Giovanni Filippini[1], Doctor Giovanni Pezzotti[1], Doctor Massimo Biagetti[1]
[1]Istituto Zooprofilattico Sperimentale dell'Umbria e delle Marche "Togo Rosati", Perugia, Italy, [2]A.N.A.B.I.C. Associazione Nazionale Allevatori Bovini Italiani Carne, Perugia, Italy, [3]Università della Tuscia (Viterbo) Laboratorio di Genetica Molecolare (DIBAF), Viterbo, Italy

Paratuberculosis (PTB), due to Mycobacterium avium ssp. paratuberculosis (MAP), is a conditioned disease. The progression of the infection depends on the containment action of innate and cell-mediated immunity (CMI), and it is related to genetic and environmental factors. In particular, PTB susceptibility seems to be associated to some genes coding for immuno-regulators involved in the cell-mediated response during infection.

This preliminary study was performed to verify an association with single nucleotide polymorphisms (SNPs) in candidate genes encoding for cytokines and their receptors, in the Italian ancient beef breed, known as Marchigiana, bred in Central Italy.

The final aim of our research, based on a longitudinal study, is the identification of phenotypic and genetic profiles characteristic of those subjects potentially able to contrast or contain MAP infection.

A MAP infected herd was subjected to traditional PTB diagnostic tests such as ELISA, qPCR and bacteriological isolation, in order to evaluate the state of MAP infection of single subjects. For that concerns CMI, Gamma Interferon (IFN-γ) test was performed to detect cytokine produced by T lymphocytes of examined animals, in response to stimulation with purified protein derivatives from Mycobacterium bovis (PPDB), Mycobacterium avium (PPDA) and MAP (PPDJ). In this way, MAP infected or exposed subjects were highlighted.

Considering the obtained results, animals were divided on the basis of their different phenotypic profiles.

DNA samples extracted from blood of 112 bovines, with a defined MAP infection status, were analyzed to verify an association with SNPs in genes encoding IFN-γ (BoIFNG), interleukin receptor 10 (IL10RA), interleukin receptor 12 (IL12RB2), Tool Like Receptors (TLR1, TLR2, TLR4).

In our population we found genotypes mostly associated to resistance pointing out a preliminary good concordance between phenotypic and genotypic profiling to be further investigated.

227

Humoral immune response against Mycobacterium avium subspecies paratuberculosis immunogenic epitopes in Type 1 Diabetes pediatric patients

Dr Marta Noli[1], Dr Gian Franco Meloni[1], Dr Davide Cossu[1], Dr Sepideh Hossein[1], Dr Somaye Jasemi, **Prof Leonardo A. Sechi[1]**

[1]*Unversity of Sassari, Sassari, Italy*

Introduction: Type 1 diabetes (T1D) is an autoimmune disorder, characterized by the production of autoantibodies to the pancreas beta-cells. In previous studies, we have indicated that Mycobacterium avium subspecies paratuberculosis (MAP) could be a potential risk factor for T1D, suggesting that MAP infection can induce to an immune imbalance. In the present study, we investigated the change over time of T1D patients humoral response against MAP antigens in comparison to healthy controls (HCs).

Materials and methods: Three different immunogenic MAP peptides homologous to host beta cell antigens were used to test sera reactivity: MAP3865c125-133, MAP3865c133-141 and MAP1,4-α157-173. Plasma samples collected from 78 patients with T1D, of which 26 at T1D onset, 27 from T1D diagnosed 1-5 years, 21 with T1D diagnosed 6-12 years, and 28 HCs were analyzed by indirect ELISA check the presence of specific antibodies.

Results: To investigate the potential differences among anti MAP antibodies in T1D onset, T1D 1-5 and T1D 6-12 patients, we performed a Kruskal-Wallis test; A statistical significative different response was observed when MAP 3865c125-133 peptide was used between patients at T1D Onset (TD1O) vs. HCs ($p < 0{,}0001$) and TD1O vs. 6-12 ($p=0{,}0416$); for MAP3865c133-141 TD1O vs. HCs ($p= 0{,}0030$) and for MAP1,4-α157-173 between TD1O vs. T1D patients 1-5 y ($p=0{,}0271$) and TD1O vs. T1D 6-12 y ($p=0{,}0062$);

Conclusion: Our data showed that the MAP epitopes were highly recognized by a specific humoral response in T1D patients compared to HCs. A high statistically significant difference has been observed between HCs compared with T1D onset and in patients with T1D over the years, showing a decrease in antibody titre after one year from onset.

240

Clinical trial (phase II) of vaccination with a heat-killed local strain of Mycobacterium avium subsp. paratuberculosis

Dr Roberto Damian Moyano[1], Dra. Maria Alejandra Colombatti Olivieri[1], Dr. Ariel Nagel[1], Med. Vet. Ximena Cuerda[1], Dra. Maria Jose Gravisaco[1], Dra. Fiorella Alvarado Pinedo[2], Med. Vet. Pedro Sosa[2], Dr. Gabriel Eduardo Traveria[2], Dra. Maria Natalia Alonso[1], Dra. Maria Isabel Romano[1]

[1]Instituto de Agrobiotecnología y Biología Molecular (IABIMO) INTA-CONICET, Hurlingham, Argentina, [2]Centro de diagnostico veterinario (CEDIVE) Facultad de Ciencias Veterinarias, Universidad Nacional de La Plata, Chascomus, Argentina

Paratuberculosis (PTB) is a chronic disease caused by Mycobacterium avium subsp. paratuberculosis and causes serious economic losses. Its control and development of an effective vaccine are essential. The aim of this study is the evaluation of the immune response induced by a virulent local strain inactivated vaccine and the evaluation of its interference with bovine tuberculosis diagnosis in a herd with a high prevalence of PTB.
We used three groups of 46 calves (1-2 months of age): 1) Silirum vaccine, 2) local strain inactivated with adjuvant Montanide ISA 201 (Seppic SA) and 3) unvaccinated group. Serum samples were taken at 2, 4-5, 10-12, and 22 months post-vaccination (mpv) for the measurement of antibodies. At 4-5 mpv whole blood was extracted to evaluate lymphocyte populations (CD4, CD8, WC1, CD335, co-marked with CD25 and IFNg) and an in-vitro infection of PBMC as performed (microbicide assay). In addition, the intradermal reaction with PPDa and PPDb were evaluated.
The group vaccinated with the local strain presented a higher and earlier (IgG) humoral immune response (with higher antibody titre at 4-5 mpv) and it was the only group that presented a significant response, with production of IFNg, from CD4, CD8, WC1 and NK cells (CD335). In the microbicide assay, no significant differences were observed. Vaccination did not yield false positive results for the official diagnosis (skin test) of bovine tuberculosis after 12 mpv.
The results indicate that the vaccine with the inactivated local strain induces a humoral and cellular immune response, slightly higher than that induced by the commercial vaccine.

243

Evaluation of a direct multiplex qPCR assay and ELISA for the diagnosis of bovine paratuberculosis in a tuberculosis endemic area

Señor Francisco Jurado-Martos[1], Sra Inmaculada Luque[2], Sra Carmen Tarradas[2], Sra Belen Huerta[2], Señor Fernando Cardoso[1]

[1]Departament I+D+i, CICAP, Pozoblanco 14400, Spain, [2]Departament of Animal Health, Faculty of Veterinary Medicine, University of Cordoba 14071, Spain

Paratuberculosis (bPTB) and tuberculosis (bTB) are the two most important chronic diseases of livestock caused by Mycobaterium spp. Interferences in the diagnosis of bTB have been attributed to Mycobacterium avium subspecies paratuberculosis infection, however, the interference of bTB in the diagnosis of bPTB has been little studied, therefore, the aim was to evaluate the possible interference of bTB in the diagnosis of bPTB.

Samples of 228 animals in a tuberculosis endemic area were collected for the diagnosis of bTB (retropharyngeal and tracheobronchial lymph nodes) and bPTB (ileocecal valve and ileocecal lymph node) by microbiological culture (gold standard tests). Furthermore, the accuracy of an indirect bPTB ELISA and direct multiplex qPCR (IS900-F57), were tested on blood serum, and on tissues (ileocecal valve and ileocecal lymph nodes) and stool, respectively

A total of 59 (25.9%) animals were positive for bTB and 53 (23,2%) were positive for bPTB. The sensitivity (SE) and specificity (SP) value of direct qPCR on tissue was of 91.7% and 59.6% for bTB + animals, whereas SE and SP were 75.6% and 88.3% respectively for bTB - animals. The SE and SP of the direct qPCR on faeces was 75.0% and 91.5% for the bTB + animals, and SE and SP of 41.5% and 97.7% for bTB - animals. Finally, the SE and SP of the ELISA for bTB + animals were 91.7% and 68.1% respectively, and 24.4% and 96.1% for bTB - animals. Higher SE values of bPTB were obtained with direct qPCR and ELISA in bTB + animals ($p<0.05$). These results suggest the need to develop new diagnostic techniques that allow monitoring of paratuberculosis in endemic or bovine tuberculosis-free areas.

In conclusion, bTB interferes with the accuracy of bPTB diagnostic techniques. These results should be taken into account when implementing control programs against bovine paratuberculosis disease.

245

Association between bovine MHC II haplotypes and susceptibility traits to bovine Johne's disease

Vmd Anabel Garcia[1], Dr. Richard Whittington[1], Dr. Aaron Thomas[2], Dr. Christopher Davies[2], Dr Jenny-Ann Toribio[1], Dr Karren Plain[1]

[1]*The University Of Sydney, Sydney, Australia,* [2]*Utah State University, Logan, USA*

In this study, we aim to identify associations between bovine MHC II haplotypes and susceptibility traits to paratuberculosis.

A total of 896 blood samples from cattle participating in an ongoing field assessment of the efficacy of Silirum® (Bovine Johne's Vaccine) were collected for sequencing and MHC II haplotype analysis. Data regarding shedding of MAP organisms (faecal culture), humoral and cell-mediated immune responses and lactation were collected longitudinally for 7-years, including vaccinated and unvaccinated cohorts. MAP infection susceptibility traits were categorised using different infection and exposure criteria. A range of test results at more than one time point/infection stage was applied. A genetic association analysis will be performed to investigate whether any of the 23 MHC II haplotypes identified were associated with MAP susceptibility traits in the population.

The pathogenesis of Bovine Johne's disease (BJD), caused by Mycobacterium avium subsp. paratuberculosis (MAP), is not completely understood. It is widely described that different disease states are possible in BJD, due to the host immune response which is directly linked to the level of exposure to the pathogen through MAP shedders on farms. The Major Histocompatibility Complex II (MHC II) is a candidate determinant of disease resistance and susceptibility because of its role in specifying antigen-presenting glycoproteins, but there is little information on this host genetic factor in bovine paratuberculosis. This study will provide a detailed description of MAP susceptibility traits and their association with bovine MHC II haplotypes.

257

Probiotic bacteria modulate immune responses to paratuberculosis vaccination

Ms Maddi Oyanguren[1], Ms Elena Molina[1], Ms Maitane Mujika[1], Ms Ainara Badiola[1], Ms Iraia Ladero[1], Ms Natalia Elguezabal[1]

[1]*Department of Animal Health, NEIKER- Basque Institute for Agricultural Research and Development-BRTA, Derio, Bizkaia, Spain*

A great amount of studies performed both in animals and humans concur that gut microbiota plays a fundamental role in immunity and it can be determinant for the efficacy of vaccination. Probiotics can exert effects on the microbiota and they have been proposed as a means of enhancing vaccine-specific immunity, although results have been inconsistent. In this work, we have investigated the effect of a probiotic in combination with vaccination on Map challenge in the rabbit model.

For this purpose the following treatments were administered previous to oral challenge with Map: a commercial inactivated Map vaccine applied alone, the same vaccine in combination with the probiotic, the probiotic alone and a non-treated group. Weight loss, bacterial burden and pathological lesions in tissues were the parameters selected to evaluate efficacy, whereas phagocyte reactive oxygen species (ROS) activity, neutrophil bacterial killing assays, antibody levels and lymphocyte subpopulation analysis were performed to evaluate immunomodulation.

Peripheral neutrophils from the only vaccinated group showed higher ROS activity against Map antigens and higher bactericidal activity, compared to the rest of the groups at 30 days post vaccination. These effects were maintained as a tendency 60 and 90 days post vaccination. The combined treatment of vaccine and probiotic presented higher plasma IgA levels and lower bacterial burden as seen by tissue culture. In the vaccinated and the combined treatment groups, Map was contained to sacculus rotundus and vermiform appendix presenting 50% and 75% reductions compared to the infected control group, respectively. These results indicate that the probiotic administered to the vaccinated animals was able to modulate the immune response, ultimately affecting protection. This work shows that the administration of a probiotic can affect the efficacy of vaccination, further highlighting the importance of intestinal bacteria in PTB immunity.

258

Optimization and Validation of Lateral Flow Assay for Rapid Diagnosis of Paratuberculosis

Dr Mukta Jain[1], Dr Manju Singh[1], Mr Amit Kumar Singh[1], Dr Saurabh Gupta[2], Dr Rathnagiri Polavarapu[3], Dr Amit Kumar[4], Professor G K Aseri[1], Dr Deepansh Sharma[1], **Dr Jagdip Singh Sohal[1]**

[1]Amity Center For Mycobacterial Disease Research, Amity University Rajasthan, Jaipur, India, [2]GLA University, Mathura, India, [3]Genomix Molecular Diagnostics Pvt. Ltd., Hyderabad, India , [4]Sardar Vallabh Bhai Patel University of Agriculture & Technology, Meerut, India

Globally livestock industry is a major employer therefore protecting animal health and production becomes priority for stakeholders. Country has a growing demand for low-price, rapid and trustworthy diagnostics in veterinary medicine. Point of care diagnostics (PoCD) have remarkable advantages over existing laboratory-based tests, owing to low cost, onsite application and rapidity. PoCD is especially important to the context of resource poor/ limited areas. Present study reports the development of PoCD lateral flow assay-based antibody detection for rapid diagnosis of paratuberculosis.

Different antigens; commercial PPA, native PPA (prepared using native strain), and specific recombinant protein (RP) mix (1693c, 2168c, 2677c, 3547c and 4308c) were evaluated as candidates to coat LFA strips. On reference positive serum (60), RP based LFA showed maximum detection limit (90.0%). Hence RP based LFA was selected for further studies. Optimized LFA had sensitivity of 77.7%, 75.4% and 75.3% with tissue culture, fecal PCR and plate ELISA, respectively. This LFA showed 100% specificity with all reference tests. In relation to plate ELISA (titer detection limit); RP based LFA had 100% sensitivity of detection when S/P ratio of serum sample is >1.0 and sensitivity is 80% in the S/P ratio range of 0.8-1.0. Positive predicted value (PPV) for this LFA was 100% with all reference tests. Negative predicted value (NPV) ranged between 73.1% to 76.3%. Statistically LFA had nearly perfect agreement with tissue culture, fecal PCR and plate ELISA. Field sera (608) were also tested using LFA and 283 (46.5%) were found positive. Higher positivity was found because these samples belonged to farms with history of clinical disease. Using present LFA diagnosis can be done in less than five minutes time. In third party validation LFA was reproducible. LFA was also reproducible using different batches of RPs. In conclusion, present LFA will play major role in controlling paratuberculosis.

259

Recombinant Johnin (rJohnin) Assay for the DTH based Specific Diagnosis of Paratuberculosis

Dr Mukta Jain[1], Mr Amit Kumar[1], Dr Manju Singh[1], Dr Saurabh Gupta[2], Dr Rathnagiri Polavarapu[3], Dr Vinod Kumar Singh[4], Dr Amit Kumar[5], Dr Deepansh Sharma[1], Professor G K Aseri[1], **Dr Jagdip Singh Sohal[1]**

[1]*Amity Center For Mycobacterial Disease Research, Amity University Rajasthan, Jaipur, India,* [2]*GLA University, Mathura, India,* [3]*Genomix Molecular Diagnostics Pvt. Ltd., Hyderabad, India,* [4]*College of Veterinary Science and Animal Husbandry, Mathura, India,* [5]*Sardar Vallabh Bhai Patel University of Agriculture & Technology, Meerut, India*

Delayed type hypersensitivity (DTH) skin testing using Johnin reagent remains one the most popular tool to diagnose paratuberculosis. This assay is useful in diagnosing the early stages of infection. However, Johnin reagent is prepared from the secretory proteins of whole cell extracts and contains many proteins shared with environmental bacteria. Therefore, specificity of this assay is compromised especially in disease endemic areas. Present study explored the potential of specific recombinant secretory proteins (1693c, 2168c, 2677c, 3547c and 4308c) as DTH markers for paratuberculosis. This assay was named as recombinant Johnin (rJohnin) as contained recombinant proteins (RPs). Initially these recombinant antigens were tested and established as cell mediated immunity (CMI) and DTH markers in guinea pig model.

Then a sheep flock and cattle herd endemic for paratuberculosis were selected to determine the positive cutoff of rJohnin assay. Animals were inoculated subcutaneously with 500 µg RPs (as per OIE suggested protein concentration) mix (100 µg per protein) as well as with traditional Johnin (as per manufacture instruction). Total 170 Animals were tested and of these 78 (45.8%) were positive in traditional Johnin and their corresponding thickness in rJohnin was ≥2.0 mm. Hence, this was decided as positive cutoff (≥2.0 mm) for rJohnin. These animals were also tested by plate ELISA and fecal PCR. Statistical analysis revealed 70.5% sensitivity and 89.7% specificity for recombinant Johnin in relation to plate ELISA. Poor correlation was found with fecal PCR. Previous studies also support this non-significant relation. Further, 231 animals (cattle, goat, sheep and buffalo) were tested from other endemic herds and 73 (31.6%) were found positive using rJohnin assay. Results of rJohnin assay are encouraging and further testing of this assay in different settings (herds with different management, disease status etc) and geographical regions will evaluate the practical applicability.

260

Penside Plate rELISA as 'Herd Screening Assay' for Onsite Diagnosis of Paratuberculosis

Dr Mukta Jain[1], Mr Amit Kumar Singh[1], Dr Saurabh Gupta[2], Dr Rathnagiri Polavarapu[3], Dr Amit Kumar[4], Dr Deepansh Sharma[1], Professor G K Aseri[1], **Dr Jagdip Singh Sohal[1]**

[1]Amity Center For Mycobacterial Disease Research, Amity University Rajasthan, Jaipur, India, [2]GLA University, Mathura, India, [3]Genomix Molecular Diagnostics Pvt. Ltd., Hyderabad, India, [4]Sardar Vallabh Bhai Patel University of Agriculture & Technology, Meerut, India

Plate ELISA is a recommended as the primary diagnostic herd screening test. In resource limited areas implementing disease control is a challenge due to non-availability of diagnostic reagents. Therefore, penside tests are need of the hour. Present study optimized plate ELISA using specific recombinant proteins (1693c, 2168c, 2677c, 3547c and 4308c) with onsite applicability. This assay has been named as plate rELISA because recombinant proteins are used as antigen. AUCROC analysis of present rELISA in comparison to tissue culture and reference sera (positive and negative) revealed ≥0.6 S/P ratio as positive cut-off. At this cut-off ROC value and Youden index was 0.95 and 0.91, respectively. rELISA showed sensitivity of 91.08% and specificity of 100% with fecal PCR. rELISA was reproducible as revealed by inter-assay CV (<15) and intra-assay CV (<10). Further, 1226 serum samples from cattle, goat, sheep, and buffalo were tested using rELISA and 326 (26.5%) were found positive. After lab optimization performance was of rELISA was evaluated for onsite applicability using a pocket size and solar chargeable battery-operated ELISA reader. Three farms of domestic ruminants having known history of paratuberculosis were selected. At these farms 240 animals were randomly selected and blood was collected using lancet. Then onsite plate ELISA was performed using these serum samples. All incubations were done at ambient temperature. OD values were read using potable ELISA reader and converted to S/P ratios. Part of the serum samples was also brought to laboratory and were again tested using laboratory-based ELISA reader. An absolute agreement was obtained between penside onsite plate rELISA and laboratory-based plate rELISA. In both assays, same 135 animals were found positive. Findings of the present study are encouraging in making plate rELISA a test that is a point of care diagnostic.

263

A team-based approach to farmer engagement in Johne's disease control

Ms Lorna Citer[1], **Lawrence Gavey[1]**, Dr David Graham[1], Mr Eamonn Farrell[2], Mr Shane McElroy[3]
[1]Animal Health Ireland, Carrick on Shannon, Ireland, [2]Irish Co-Operative Organisation Society, Dublin, Ireland, [3]Glanbia Ireland, Kilkenny, Ireland

One of the challenges of Johne's control is motivating farmers to modify existing animal health management practices.
During a pilot programme, prior to the Irish Johne's Control Programme (IJCP), multiple stakeholders provided information to farmers. A survey, post-pilot phase, identified the potential for mixed messaging and the importance of clearly communicating key management actions and the benefits of these. Subsequently, stakeholders requested Animal Health Ireland (AHI) develop a farmer-friendly Farmer Awareness Seminar (FAS) which could be readily delivered as part of the IJCP.
Co-operatives are one of the principal beneficiaries of the IJCP and have a strong presence in the Irish dairy industry with field staff frequently interacting farmers. They became the lead agencies for delivery of the seminars in conjunction with Approved Veterinary Practitioners (AVPs) as technical experts.
Key messages focusing on hygienic calf-rearing practices and farm biosecurity were developed to ensure a consistent story across all stakeholders. These formed the basis for an interactive seminar which incorporated adult-learning principles and strategies to empower farmers to implement change. Support materials including a facilitators' guide and supplementary activity sheets, were also developed.
A series of 'train the trainer' workshops were held to upskill field staff and to ensure a common understanding about Johne's disease management as well as the operational aspects of the IJCP. The same communication messages were provided during training workshops for AVPs.
Twelve co-operatives presented 43 Farmer Awareness Seminars with 633 farmers attending over a five-month period. The evaluation process required farmers to rank their experiences and included an option for free text responses.
The FAS generated a high level of satisfaction with farmers and field staff positively commenting on the interactive approach. Consequently, some farmers indicated their willingness to make management changes to control Johne's and expressed an interest in Phase Two of the IJCP.

264

Direct qPCR is equally sensitive as liquid culture for the detection of MAP in ovine faecal samples

Dr. Heike Koehler[1], Chris Pickrodt[1], Dr. Petra Möbius[1]
[1]*Friedrich-Loeffler-Institut, Federal Research Institute for Animal Health, Jena, Germany*

Cultural detection of MAP in ovine faeces and tissues is laborious, because sheep are often infected by MAP S-type strains, which grow even more slowly than MAP C-type strains. Liquid culture using a modified Middlebrook 7H9 medium, which is more sensitive than culture on solid media, takes up to 12 weeks for reliable results. Direct detection of MAP genome in clinical samples by qPCR has the potential to speed up diagnostics considerably. Aim of the present study was to compare the results of liquid culture and direct qPCR for the detection of MAP in faecal and tissue samples of German sheep.
Eighty faecal and 13 tissue samples (small intestine or mesenterial lymph nodes) of individual sheep with previously undefined MAP status were analysed in parallel. The samples had been submitted to the German National Reference Laboratory for paratuberculosis in the context of flock monitoring programmes of regional animal health services. Liquid culture was performed for 12 weeks after decontamination with HPC using a modified Middlebrook 7H9 medium supplemented with egg yolk and mycobactin J. Presence of MAP in the culture broth was proven after magnetic separation by endpoint PCR targeting IS900. Direct detection of MAP genome in the faecal and tissue samples was done using a commercial qPCR kit (Adiavet ParaTB real time, Bio-X) after DNA extraction using the QIAamp DNA Mini Kit (Qiagen).
MAP was detected in 18/93 samples by direct qPCR and in 16/93 samples by liquid culture. The accordance between both methods was 91.4 % with matching results for 13 positive and 72 negative samples. These data confirm that direct detection of MAP genome by qPCR is a reliable and less laborious alternative to cultural isolation allowing rapid detection of MAP in faecal and tissue samples of sheep.

265

Exploring the best peptide regions of secretome of Mycobacterium avium subspecies paratuberculosis for better functionality and binding efficacy to design peptides based ELISA

Dr Sujata Hariharan[1], Dr. Kundan Chaubey[1], Dr. Pravin Singh[1], Dr. Shoor Vir Singh[1]

[1]*Gla University, Mathura, Institute Of Applied Sciences And Humanities, Gla University, Mathura (u.p.) - 281406, India, India*

Background:Johne's disease, a chronic enteric disease of ruminants is caused by Mycobacterium aviumsubspecies paratuberculosis (MAP) and is responsible for huge economic losses to the livestock industry globally. Control and eradication of Johne's disease is considered difficult because of its insidious nature, long incubation period and inability to detect the disease early and accurate. Thus, there is scope to develop more specific diagnostic assay. Here, we studied the potential peptide region of MAP secretory antigens to develop more specific diagnostic assay for paratuberculosis.

Methods: The peptide sequences of selected MAP specific proteins were subjected to MHC-I Binding Predictions (IEDB analysis resource) to identify the putative regions that are capable of interacting with the bovine lymphocytic antigen-T7 (BOLA-T7). Based on the NetMHCpan server, the respective secretome sequences were predicted for their functional regions vis-à-vis interaction to MHC-I based on the MHC class I pseudo-sequence, neural network training, pseudo-distance and nearest neighbor based algorithms applied. This exercise generated a set of amino-acid sequences ranked based on functionality with respect to interaction with the BOLA-T7 for each MAP secretome.

Results: Study identified a few active peptides of MAP secretome that had binding affinity with BOLA-T7 at defined parameters. The level of affinity between identified peptides and BOLA-T7 indicates its biological plausibility and its potential as suitable candidate for laboratory validation.

Conclusions: In-silicostudygives ample choice for selection based on the best peptide region of the secretome with good functionality and binding efficacy. Thepeptide sequence identified in this study has diagnostic potential, thus needs to be validated in laboratory.

268

No evidence of transmission of Mycobacterium avium subspecies paratuberculosis from the udder skin into the colostrum of dairy goats

Miss Chris Pickrodt[1], Dr. Karsten Donat[2], Dr. Udo Moog[2], Dr. Heike Köhler[1]

[1]*Institute of Molecular Pathogenesis, Friedrich-Loeffler-Institut (Federal Research Institute for Animal Health), Jena, Germany,* [2]*Thuringian Animal Diseases Fund, Animal Health Service, Jena, Germany*

Colostrum is considered a source for early infection with Mycobacterium avium subspecies paratuberculosis (MAP). MAP is generally assumed to be secreted directly into the colostrum and/or introduced by faecal contamination during milking or storage before feeding the kids. This study aimed at determining whether a disinfection of the udder can lead to a reduction of the MAP burden. Colostrum samples of 55 goats from a paratuberculosis affected herd with high contamination of the bedding in the lambing pen and the waiting pen as determined by environmental sampling were tested for the presence of MAP. Nine goats (16.4%) were confirmed as MAP positive before the study by faecal culture.

One colostrum sample per goat was obtained before and one after treatment of the udder skin with a mycobactericidal disinfectant on the day of lambing. Besides, bulk colostrum samples were collected. Before disinfection a swab sample from the udder skin was taken with a dry cotton swab to quantify the MAP burden. Individual and bulk colostrum samples were analysed by immunomagnetic separation followed by qPCR using commercial kits. DNA extraction and qPCR as well as cultural isolation on Herrold's Egg Yolk Medium were performed for swab samples.

The limit of detection of the methods as evaluated before was between 10^1 and 10^2 cfu/mL for colostrum samples and 10^3 cfu/cm^2 for swab samples. None out of 110 individual and 14 bulk colostrum samples showed a positive result. MAP was detected in 1/27 udder swabs by qPCR but not cultivation.

Despite the remarkable MAP contamination of the barn environment, MAP was not detected in most samples and therefore, a substantial transmission of the pathogen via the udder skin into the colostrum during the milking process is unlikely. In MAP affected goat herds hygienic milking without additional udder skin disinfection can be recommended.

269

Evaluation of four tests (screening and confirmatory) for the presence of MAP infection in the spiked and un-spiked milk samples

Dr Kundan Kumar Chaubey[1,4], Dr Manju Singh[2], Dr Saurabh Gupta[1,4], Dr P K Singh[1], Dr A V Singh[3], Dr Dev Chauhan[3], Dr Shoor Vir Singh[1,4]

[1]*Department of Biotechnology, GLA University, Mathura, Mathura, India,* [2]*Adesh Institute of Medical Sciences & Research, Adesh University, Bathinda, Bathinda, India,* [3]*ICMR-National JALMA Institute for Leprosy and Other Mycobacterial Diseases, Agra, Agra, India,* [4]*ICAR-Central Institute for Research on Goats, Makhdoom, Mathura, India*

Background: Mycobcaterium avium subspecies paratuberculosis (MAP) infection leads to major production losses in domestic livestock. Infection of humans through consumption of milk and milk products made from pasteurized milk is not safe and is serious public health concern globally, since MAP is not inactivated during pasteurization. Microscopy (ZN-staining), Indirect fluorescent antibody test (i_FAT), IS900 PCR and IS1311 PCR} were evaluated for detection of MAP in spiked and un-spiked milk samples.

Method: Samples (104) of raw and pasteurized liquid milk were spiked and un-spiked with MAP. Milk samples were heated to 80.00C before screening by Microscopy, i_FAT, IS900 PCR and IS1311 PCR.

Results: Of 104 milk samples (raw and pasteurized), 26 were spiked with MAP and rest 74 were un-spiked. Of these 26 spiked milk samples, 24 (92.3%), 12 (46.1%), 24 (92.3%) and 24 (92.3%), were detected positive by microscopy, i_FAT, IS900 PCR and IS1311 PCR, respectively. Sensitivity of microscopy, IS900 and IS1311 PCR was high (92.3%) followed by i_FAT (46.1%). However, percent infection of both raw and pasteurized milk with MAP infection was very high (24.0 to 37.0%) in India. Lower detection of MAP bacilli by i_FAT may be due to curdling of milk samples during pasteurization, since due hot summer season at that time and un-proper storage during transport led to capturing the bacilli by curdled milk. Results showed that when infection of MAP is high (here due to spiking), any of the two tests combinations (Microscopy with IS900 or IS1311 PCR), may be chosen as the tests of choice for the detection of MAP infection.

Conclusions: Sensitivity and specificity of three tests was equal (92.3%), providing equal opportunity for detection in spiked or during heavy contamination by MAP. However, sensitive of the i_FAT test was poor in pasteurized milk samples, therefore may not be optimum test.

274

Understanding the antibacterial mechanisms of copper ion treatment on Mycobacterium avium subsp. paratuberculosis

Mr Carlos Tejeda[1], Mss Marcela Villegas[1], Dr Pamela Steuer[1], Dr Miguel Salgado[1]

[1]Instituto de Medicina Preventiva Veterinaria, Facultad De Ciencias Veterinarias, Universidad Austral De Chile, Valdivia, Chile

Introduction: Copper and its alloys are natural and very well-proven antimicrobial materials. The mechanisms of action through which copper is highly effective to have been described at the molecular and cellular levels. However, both the design of the studies carried out and the nature of the microorganisms studied have meant that this research has been of limited scope. In the present study, we examined the action mechanisms of a copper ion treatment on the integrity of Mycobacterium avium subsp. paratuberculosis (MAP), a highly resistant animal pathogen.

Material and Methods: To investigate the mechanisms that would explain how copper ions exert effective control on MAP, a study was proposed that evaluated the integrity of the nucleic acids (Comet assay), proteins (ROS and oxidation proteins), and cell wall (lipid oxidation) of this pathogen, as response variables against this treatment.

Results: The copper ion treatment applied to MAP cells resulted in nucleic acid degradation and disintegration, increased ROS production, and protein alteration. However, it had no effect on the integrity of the MAP cell wall, which offers a possible scientific explanation for the observed tolerance of this pathogen to this copper-based treatment.

Conclusion: This new evidence about the observed tolerance in the MAP cell wall against the copper ion may help us to understand how we can improve the proposed copper-based treatment, and finally achieve a totally effective alternative to control MAP.

275

A novel paraffin slide culture (PSC) technique to isolate and characterize Non-tuberculosis mycobacteria (NTM)

Dr Mukta Jain[5], Ms Simpi Maurya[1], Dr Rahul Narang[3], Dr Pratibha Narang[3], Ms Sakshi Rana[1], Dr Amit Kumar[4], Dr Gajendra Aseri[1], Dr Deepansh Sharma[1], Dr Jagdip Singh[1]

[1]Amity University, Rajasthan, Jaipur, India, [2]E-YUVA, Centre, Chandigarh, India, [3]Mahatama Gandhi Institute of Medical Sciences, Sevagram Maharashtra, India, Vardha, India, [4]Sardar Vallabh Bhai Patel University of Agriculture & Technology, Meerut Uttar Pradesh, India, Meerut, India, [5]Biogenix Inc. Pvt. Ltd., Lucknow, India

Nontuberculous mycobacteria (NTM) are well established to cause lung and other diseases. Recently, pulmonary NTM infection have dramatically increased globally. Due to differences in treatment regime discrimination of NTM from Mycobacterium tuberculosis (MTB) complex is critical. Sputum microscopy cannot discriminate between NTM and MTB complex. Thus, NTM infections are commonly misdiagnosed as pulmonary TB, especially in resources-limited areas. Presently, WHO recommends immediate anti-TB treatment after smear-positive results. Hence NTM infection is not considered and may result in inappropriate treatment. Culture followed by biochemical/molecular identification is more definitive method. Commercially paraffin slide culture (PSC) technique is available that supports the growth of NTM but not of MTB complex. Mahatama Gandhi Institute of Medical Sciences (MGIMS), Sevagram developed an inhouse method of PSC in early 21st century and found PSC a better tool to isolate NTMs compared to other methods. However, PSC is not popular for isolation of NTM. Present study evaluated the utility of PSC in isolation and characterization of NTM. Inhouse PSC method optimized by MGIMS, Sevagram was adopted. Briefly, paraffin was sterilized by autoclaving and coated on sterile glass slides. These slides were placed in McCartney bottles containing 5.0 mL of basal salt solution/ Czepak broth containing PANTA antibiotic mix. These bottles were inoculated with 0.5 mL decontaminated inoculum from soil and water. Of the total 40 samples inoculated (soil-16 and water-24), 12 were positive (soil-08 and water-04) for NTM. Colonies appearing on paraffin slides were directly submitted to ZN staining and PCR. It took 20-30 days for colonies to appear. NTM nature was confirmed due to presence of hsp65 gene and absence IS6110 element. For speciation biochemical analysis and/or restriction analysis (BstEII and HaeIII) of hsp65 PCR product can be done. In conclusion, inhouse PSC technique was found user-friendly tool to isolate NTM.

277

Search for bacterial gut microbiome patterns and factors associated with pathogenesis of Paratuberculosis in goats

Dr. Petra Moebius[1], Prof. Elisabeth Liebler-Tenorio[1], Dr. Heike Koehler[1], Chris Pickrodt[1], Prof. Thilo Fuchs[1]
[1]Institute of Molecular Pathogenesis, Friedrich-Loeffler-Institut (Federal Reseach Institute for Animal Health), Jena, Germany

Studies on the intestinal microbiome during MAP infection and the influence of the indigenous animal microorganisms on the development of chronic paratuberculosis (Ptb) are still rare.
The study presented here is focused on goats and aims to identify patterns and factors potentially involved in pathogenesis of this disease, thus revealing a possible association between Ptb and the composition and function of the bacterial microbiome in the gut.
Goats of the breed "Thueringer Wald-Ziege" originating from goat farms with and without a history of Ptb were sampled. Beside a longitudinal study regarding the faecal microbiota of healthy goats of different ages, samples of content from seven intestinal compartments of clinically diseased animals were examined by 16S rRNA sequencing compared with those of the healthy animals. The health status of autopsied animals was checked by macroscopy, histology and microbiology.
Preliminary results show a higher α-diversity in healthy than in clinically diseased goats in the content of small as well as large intestine. The β-diversity was also different between samples from healthy and clinically diseased goats in these both compartments, and depending on age and farm origin. The altered bacterial microbiota in the small intestine of individual clinically diseased goats show an emergence of Peptostreptococcaceae, Enterobacteriaceae, Chlostridiaceae and a clear reduction of Christensenellaceae, Atopobiaceae, Oscillospiraceae, Ruminicoccaceae and others in comparison to not infected goats. In faecal microbiota of clinically diseased goats, changes were found in 13 bacterial families compared to healthy goats, but with very low relative abundance.
A larger panel of samples will be investigated in a similar manner, and additional functional analyses will be performed to determine relevant microbial activities in healthy and diseased animals, which play a role in the onset and progression of the disease.

278

Development of a diagnostic test to identify MAP in the blood and tissue of patients with Crohn's disease

Snr Research Fellow Neil Rayment[1], Dr Amy Hermon- Taylor[1], Mr Barry Hudspith[1], Professor Roger Pickup[2], Dr Glenn Rhodes[3], Dr Gaurav Agrawal[1], Professor Jeremy Sanderson[1], Professor John Hermon-Taylor[1]
[1]Kings College London, London, United Kingdom, [2]Faculty of health and Medicine, Lancaster University, Lancaster, United Kingdom, [3]Centre for Ecology &Hydrology, Lancaster environment centre, Lancaster, United Kingdom

We have developed, and are in the process of validating, a panel of monoclonal antibodies which will be used to provide a simple diagnostic test for MAP infection in humans. The targets of these antibodies are based on the recognition of a specific abundantly expressed cell surface protein (a product of the IS900 gene) comprising two transmembrane regions with extracellular amino and carboxy terminal domains. Antibodies directed to the amino terminal domain are designated A0X and A0XP and those to the carboxy terminal domain A4 and XA4P (where P denotes the phosphorylated derivatives)
Following conjugation with immunofluorescent labels, preliminary validation experiments have shown positive staining with a panel of MAP culture preparations but negative for a panel of other non – MAP mycobacteria.
In a study of human gut tissues, endoscopic biopsy and operative samples from 48 people with Crohn's disease, positive areas of fluorescence antibody staining were identified in both the mucosa as well as the lamina propria. This signal appeared occasionally as small foci but larger 'cluster like' signals were also observed. Individual labelling of antibodies showed co-localisation of signal suggesting target specificity has been achieved. Further validation experiments are on-going.
The continuing absence of a reliable practically applicable diagnostic for MAP in humans leaves the issue unresolved as it is generally considered then set aside. The ability to visualise MAP in human tissue for the first time , as shown here, may offer new insights into mechanisms of disease. These monoclonal reagents are applicable to tissues, to circulating naturally infected white blood cells and to experimentally infected cells. This methodology can easily operate within existing NHS blood Flow Cytometry facilities and in routine paraffin embedded histopathology formats. It is equally applicable as a veterinary diagnostic.

279

Developing a novel Johne's Progress Tracker to support the UK National Management Plan for Dairy herds

Mr Pete Orpin[1,2], Mr Nick Taylor[3], Mr Dick Sibley[1,4]

[1]Myhealthyherd, Leicester, United Kingdom, [2]Special Lecturer Nottingham University, Nottingham, United Kingdom, [3]VEERU, Reading University, Reading, United Kingdom, [4]Westridge Veterinary Group, Tiverton, United Kingdom

Introduction: he UK National Johne's Management Programme (NJMP) has engaged over 8700 dairy producers with Johne's Disease (JD) risk assessment, surveillance and vet led control plans. The predominant surveillance and control strategies involve milk ELISA testing with an estimated 71% using strategic testing, management and culling to manage the disease. The next stage with the NJMP was to develop a more robust monitor of progression utilising improved outcome measures and drivers of JD infection within herd spread.

Methodology: A pilot study from 39 herds (652 test dates, 157,985 test results) utilising quarterly milk ELISA testing was evaluated to define more robust outcomes and drivers for Johne's disease in dairy herds.
The key outcomes identified were Average Test Value of all results, % cows positive at differing thresholds (> 30,60,100). The key drivers identified were New Detection rate (index case test positive), % repeat positive and relative risks of culling and service of JD positive cows compared to non JD positive cows
The measures were subsequently benchmarked over 257 randomly chosen quarterly milk testing herds, over 7 test periods in 2018-19, to develop quartile ranges to produce a template report called the Johne's Progress Tracker.
The database code was developed within Interherd + and then shared with the three main Milk Recording Organisations.

Results: The use of the revised outcome measures and infection drivers has delivered new insights into established ELISA testing programmes. The commonest failings were failing to control infection risks and an over reliance on culling as a control point. Risk management is central to JD control.

Summary: More effective analysis and graphical illustration of outcome and driver measures of data can deliver improved JD control at farm level and is more motivational than simply benchmarking herds by % testing positive.

281

Part 2: UK approach to Johne's Disease control -Utilising web-based tools to enhance farmer engagement

Mr Pete Orpin[1,2], Mr Dick Sibley[1,3]
[1]Myhealthyherd, Cosby, United Kingdom, [2]Special Lecturer Nottingham University, Nottingham, United Kingdom, [3]Westridge Veterinary Group, Tiverton, United Kingdom

Introduction: The complexities of a long incubation period, tests of low sensitivity and discrepancies between test and true herd prevalence hamper Johne's Disease (JD) control resulting in farmer and vet despondency. Developing tools to explain risk and prevalence better has been central to the success of the UK JD program

Methodology: Two web-based tools have been the cornerstone of vet and farmer engagement with JD control.
Myhealthyherd- a web-based health planning tool based on a predict and prevent model of disease management. Risks are captured and traffic light scored to assess the likelihood of disease introduction and spread. Within the JD module a prevalence prediction graph was developed to illustrate future true herd prevalence. A control planner with inbuilt robustness checker allows for more effective JD plans to be created. Plans that were robust were highlighted green and those requiring further work amber or red.
Herdwise- A web-based reporting system for interpreting quarterly milk ELISA results using a traffic light system based on the Danish control system. Cows are categorised into high, medium and low risk and disease control progression is illustrated by graphs tracking prevalence and date of birth of infection. Cohort monitoring and latterly the integration of the Johne's Progress Tracker have further enhanced the value of the testing program.

Results: The use of graphical based web-based tools aids cost effective management of JD. The traffic light scoring of risks and results avoided the more typical binary positive/ negative reporting of results. The Myhealthyherd and Herdwise programs have become tools to help farmers understand Johne's control more readily.

Discussion: JD is a complex disease requiring risk management and behaviour change to effectively control spread. Graphical web-based tools provide a simple, cost effective way of illustrating the complexity of JD whilst maintaining farmer engagement.

283

Achieving 93% uptake with UK National Johne's Management Plan using commercial drivers and industry support

Mr Pete Orpin[1,2], Mr Dick Sibley[1,3]

[1]*Myhealthyherd, Leicester, United Kingdom,* [2]*Special Lecturer Nottingham University, Nottingham, United Kingdom,* [3]*Westridge Veterinary Group, Tiverton, United Kingdom*

Introduction: In 2009 a National Johne's engagement plan was developed with the support of milk processors, laboratories, database providers, farming and veterinary groups. By December 2021 93% (8,700) of the UK Dairy Farmers will have completed a risk assessment, surveillance and have installed a JD control plan utilising an accredited vet.

Methodology: Milk processors and retailers define the priorities for UK Dairy farmers as part of their contractual arrangements. A cohesive national plan (NJMP) for Johne's Disease (JD) control was developed in 2015 which with the objective of JD reduction in the UK.
A flexible framework for JD control was established allowing farmers to choose one of 6 potential control strategies using advice from an accredited JD vet.
The success has been achieved through focusing on creating practical JD plans that match the need of the farmer and create commercial consultancy opportunities for the private vet. The engagement of milk processors and laboratories are major drivers to engagement of farmers.

Results: In October 2019 the NJMP became a compulsory National Farm Assurance standard for all dairy farms supplying farm assured milk.
47 milk processors with an estimated 93% of UK producers are now engaged with the NJMP. By December 2020 declared control strategies from 5215 farmers were improved farm management (IFM) and strategic testing (42%), IFM and test and cull (28%), biosecurity protect and monitor (11%), breed to terminal sire (12%) IFM alone (6%) and vaccination (0.03%). Data on strategies was not collated from 11 farmers.
Over 1700 vets have been accredited by the British Cattle Veterinary Association to deliver the JD plan.

Discussion: Milk processor and Retailer influence created drivers for engagement. The development of a practical, flexible and commercial approach to JD control, which appeals to all, is central to the success of the NJMP.

288

Performance evaluation of an indirect milk-ELISA for the diagnosis of Mycobacterium avium subsp. paratuberculosis in individual samples of dairy sheep

Prof Anna-Rita Attili[1], Dr Eleonora Bonacucina[2], Dr Martina Linardi[1], Dr Luciana Pacifici[3], Prof Victor Ngu Ngwa[4], Prof Vincenzo Cuteri[1]

[1]University of Camerino, Via Circonvallazione 93/95, Italy, [2]Freelance Veterinarian , Castelraimondo, Italy, [3]Official Veterinarian, ASUR Area Vasta 3, Camerino, Italy, [4]University of Ngaoundéré, Ngaoundéré, Cameroon

The aim of this study was to evaluate the performance and the most appropriate cut-off value in dairy sheep samples of an indirect ELISA (ID Screen® Paratuberculosis - ID.vet, France). For the purpose, an intra-flock epidemiological study was carried out in Italy on 200 paired blood and milk sheep sera and 200 faecal samples collected both in October 2019 (T0: low milk production) and in April 2020 (T6: high milk production) in two flocks with (Flock-A: n=150) and without (Flock-B: n=50; Control group) history of Paratuberculosis. The population was stratified according to age: young, adult, and elderly sheep. Indirect ELISAs and qPCR IS900 were performed following the Manufacturer's instructions. As Gold standard test, blood indirect ELISA and PCR were considered. S/P blood- and milk-ELISA ratios were compared using Pearson's correlation coefficient (R2) while Simple Interactive Statistical Analysis and GraphPad-Prism-8 software were used to evaluate the performance of the indirect milk-ELISA and conduct the ROC curve analysis. A strong correlation between blood- and milk-ELISA were observed both at T0 (R2=0.8348) and at T6 (R2=0.7200). In Flock-A, 29 sheep (19.33%, n=150) resulted positive for Mycobacterium avium subsp. paratuberculosis (MAP) excretion. The strong faecal shedders (2% at T0: mean Ct=22.23; and 3.33% at T6: mean Ct=36.07) resulted positive to both blood- and milk-ELISAs in all the three age groups. In Flock-B all ewes resulted non-shedders and negative for MAP antibodies, in both blood and milk sera. The milk-ELISA obtained an 'Almost perfect agreement' (kappa=82.6%) at T0 and a 'Substantial agreement' (kappa=70.0%) at T6, suggesting a new appropriate cut-off for milk serum in dairy sheep (ROC curve analysis: >21.38 S/P% with Sensitivity>85.93% and Specificity>99.22%). Considering the new cut off, the periodic MAP indirect milk-ELISA (ID.vet, France) resulted a valid, useful alternative for MAP individual screening at different ages and stages of lactation.

292

In contrast to successfully detecting viable MAP from blood using a phage/qPCR assay within 48 hours, by two years routine mycobacterial blood culture provides no clinically useful information

Dr Robert Greenstein[1], Dr Liya Su[1], Dr Judy Stabel[2], Professor Irene Grant[3], Dr Antonio Foddai[3], Dr Sheldon Brown[1,4]

[1]Veterans Affairs Medical Center Bronx NY USA, Bronx , United States, [2]USDA-ARS-NADC, Ames, USA, [3]Queens Univerity, Belfast, United Kingdom, [4]Mt Sinai School of Medicine, New York , USA

Introduction: We have published on a MAP phage/qPCR assay, which takes ~48 hours to complete, that detects viable MAP in the blood of cattle with Johne disease (#=32) and their controls (#=17). Simultaneously, all samples were routinely submitted to conventional mycobacterial culture. We herein report on samples where colonies were detectable at ≥2 years following inoculation.

Methods: White blood cells were isolated from cattle whole blood using a Ficoll gradient, lysed, 25% of the lysate was then cultured using both Pozzato and 7H-9 media in MGIT, both of which contained PANTA. When evidence of growth was detected, samples were sub-cultured onto Herrold's egg yolk slopes. Detected colonies were sub-cultured in 7H-9 to permit nucleic acid identification. All cultures were supplemented with Mycobactin J.

Results: A total of 150 samples were received and assayed from 17 Control and 32 Johne animals. With a minimum of two years follow up, 4.0% (6/150) samples have visible colonies. Control animals have 12% (2/17) and Johne cattle have 12.5% (4/32) visible colonies. Of these six animals, MAP phage /qPCR was positive in three animals one of which was a Control animal, ELISA was positive in two animals which had Johne, IFN-γ was positive in three animals all with Johne, and fecal IS-900 PCR was positive in four animals all with Johne.

Conclusions: Clinically useful detection of viable MAP in blood is feasible within 48 hours using a rapid phage/qPCR assay. In contrast, although of intellectual and academic interest, attempting to culture mycobacteria from blood and treated as recommended with PANTA antibiotics, has no role in clinical practice. In a miniscule number of cases, it may be of use in mycobacterial nucleic acid analyses.

293

Proteomic analysis of sera from Holstein Friesian cows with different pathological forms of bovine paratuberculosis (PTB)

Miss ALEJANDRA ISABEL Navarro Leon[1], Dr. Marta Muñoz[1], Miss Natalia Iglesias[1], Miss Cristina Blanco Vázquez[1], Dr. Ana Balseiro[2], Dr. Fernando J. Corrales[3], Dr. Sergio Ciordia[3], Dr. Fátima Raquel Milhano[3], Dr. Rosa Casais[1]

[1]Center for Animal Biotechnology, Servicio Regional de Investigación y Desarrollo Agroalimentario (SERIDA), 33394 , Gijón, Spain, [2]Departamento de Sanidad Animal, Facultad de Veterinaria, Universidad de León; Instituto de Ganadería de Montaña (IGM, CSIC-ULE), Finca Marzanas, Grulleros,24346 , León, Spain, [3]Functional Proteomics Laboratory, National Center for Biotechnology, Consejo Superior de Investigaciones Científicas (CSIC), , Spain

The lack of sensitive diagnostic methods able to detect Mycobacterium avium subspecies paratuberculosis (MAP) subclinical infected animals is an obstacle for PTB control. Sera collected from naturally infected Holstein Friesian cows with different pathological forms of PTB and negative control animals with no lesions detected (n=4, each group) were analyzed using TMT-6plex quantitative proteomics to identify and quantify peptides/proteins differentially expressed (DE) between groups. Three comparisons were performed: focal versus (vs.) control, diffuse vs. control, and focal vs. diffuse. One (focal vs. control), eight (diffuse vs. control) and four (focal vs. diffuse) DE bovine proteins (q-value≤ 0.05) were identified. Alpha-1-acid glycoprotein (ORM1) was found to be upregulated in animals with focal and diffuse lesions vs. control animals, while Bovine Fetuin-B, Vitamin D-binding protein, and Complement component C6 were found exclusively upregulated in animals with diffuse lesions. Upregulation of Lipopolysaccharide-binding protein and Serpin domain-containing protein was found in animals with focal lesions vs. animals with diffuse lesions. In addition to the above-mentioned DE proteins, other proteins (q-value> 0.05 and p-value <1-π0) likely to be involved in relevant biological processes were analyzed with the Ingenuity Pathway Analysis (IPA) software tool. These proteins are involved in different biological processes such as humoral immune response, inflammatory response, and organismal injury and abnormalities. Considering the fold change, cellular localization, and biological function, eight of these DE proteins included in the resulting IPA networks were selected for further validation by specific Western Blot analysis and ELISA, using a set of serum samples from 127 animals with different pathological forms of PTB and 138 control animals from PTB-free farms. In this study, we obtained the serum proteomic profiles of animals with different pathological forms of MAP infection, which contributes to a better understanding of the pathogenesis of PTB and could provide new potential diagnostic biomarkers for PTB.

295

Evaluation of diagnostic tests for Mycobacterium avium subsp. paratuberculosis following a single screening of animals in herds endemic with Johne's disease

Mrs Elvira Ramovic[1], Mr Robert Cannon[1], Mr Dermot Yearsley[1], Mrs Margaret Griffin[1], Dr Deirdre Prendergast[1], Ms Tracy Clegg[2], Mr Darren Hand[1], Mrs Eadaoin NiGhallchoir[1], Dr Bryan Markey[3], Dr John Egan[1]
[1]Bacteriology/Parasitology, Backweston Campus, Celbridge, Co. Kildare, Ireland, [2]Centre for Veterinary Epidemiology and Risk Analysis, School of Veterinary Medicine, University College Dublin, Belfield, Dublin 4, Ireland, [3]School of Veterinary Medicine, University College Dublin, Belfield, Dublin 4, Ireland

Ten herds (9 dairy and 1 beef) in which Johne's disease (JD) was confirmed in the preceding 1 to 10 years were used to evaluate a number of diagnostic tests for the disease. A total of 1606 animals over two years of age were sampled for faeces and blood. In addition, 930 milk samples were also collected. Faecal samples were tested in liquid culture (Cornell double incubation method and Trek ESP II system), and by direct faecal PCR using a commercial real time PCR kit. Serum and milk samples were tested by MAP ELISA using a commercial screening kit and suspect / positive samples were retested using a commercial verification MAP ELISA kit. MAP was cultured from the faeces of 195 animals while PCR identified 710 animals as positive. Serum ELISA identified a low number of culture positive animals (46/195) and milk ELISA detected 38 cases out of the 117 culture positive animals from whom a milk sample was obtained. Kappa test agreement was found to be weak between culture and PCR (0.19), as well as between culture and ELISA (0.19). However, moderate agreement was recorded between milk and serum ELISA (0.47). Using liquid culture as the gold standard, the sensitivity of direct faecal PCR was calculated at 80% while specificity was 60.7%. In comparison to liquid culture, the serum ELISA method was 14.4% sensitive and 98.7% specific, while milk ELISA sensitivity and specificity was estimated at 18.8% and 98.0% respectively. In general PCR and ELISA sensitivity increased in line with increased faecal shedding of MAP. Refinement of the PCR assay cycle thresholds (Ct) for identifying JD infection may be warranted in heavily infected herds.

296

Longitudinal study on Mycobacterium avium subsp. paratuberculosis antibody kinetics and excretion in dairy sheep using blood, milk and faeces throughout the lactation period

Prof Anna-Rita Attili[1], Dr Eleonora Bonacucina[2], Dr Martina Linardi[1], Dr Luciana Pacifici[3], Dr Livio Galosi[1], Prof Victor Ngu Ngwa[4], Prof Vincenzo Cuteri[1]

[1]University of Camerino, Matelica, Italy, [2]Freelance Veterinarian , Castelraimondo, Italy, [3]Official Veterinarian, ASUR Area Vasta 3, Camerino, Italy, [4]University of Ngaoundéré, Ngaoundéré, Cameroon

The dynamics of immune responses and excretion of Mycobacterium avium subsp. paratuberculosis (MAP) during the different stages of natural infection were evaluated in an intra-flock longitudinal study among asymptomatic Italian dairy sheep. In October 2019 (T0), April 2020 (T6), and June 2021 (T20), a cross-sectional and prospective incidence studies were performed on 150 ewes. The population was stratified by age (young6 months; 6adult24 months; elderly24 months). With the exception of 50 lambs in October 2019, paired samples of blood (n=402), milk (n=352), and bulk tank milk (BTM: n=3) were collected for indirect diagnosis (ELISA IDScreen® Paratuberculosis, ID.vet, France), while faecal samples (n=402) were submitted to qPCR IS900 investigation (IDGene® Easy Preparation, and IDGene™ Paratuberculosis Duplex, ID.vet, France). Blood seroprevalences of 11.33% and 14% were detected in T0 and after 6 months (T6), respectively, with an incidence rate of 3%. Milk seroprevalences of 8% (T0) and 10.67% (T6) were observed, with an incidence rate of 7.1%. A significant increase in the mean serum antibody level was observed in the elder's ewes, both in blood (P=0.0005) and in milk (P<0.0001). At T20 incidence rates of 2.7%, 5.3%, and 4.8% were observed by blood ELISA for young, adult, and elderly ewes, respectively. Strong faecal excretion increased the incidence rate from 1.36% (n=147; T0-T6) to 3.9% (n=102; T6-T20), involving adult and elderly ewes that resulted positive (50%, n=4) to blood- and milk-ELISA at T20. The BTM analyses (S/P% at T0: 20.7% and T20: 28.4%), confirmed that MAP positivity has a seasonal trend linked to the lactation stage. Considering the relevance of MAP infection in dairy sheep and the milk processing industry in Italy, the implications for Public Health, and the need for legislation as well as in cattle, this study contributes to a better understanding of MAP spread in dairy sheep flocks.

297

Joint control of bovine tuberculosis and paratuberculosis – the case for and against

Emeritus Professor Richard Whittington[1], **Dr. Susanne Eisenberg[2]**, Dr Frank Griffin[3], Dr Anita Michel[4], Dr Ebba Schwan[5], Dr Karen Stevenson[6], Dr Maarten Webber[7]

[1]The University Of Sydney, Camden 2570, Australia, [2]Niedersächsische Tierseuchenkasse, 30169 Hannover, Germany, [3]Disease Research Limited, Invermay Agricultural Centre, Mosgiel 9092, New Zealand, [4]Faculty of Veterinary Science, University of Pretoria, Onderstepoort 0110, South Africa, [5]Swedish Farm and Animal Health, 62254 Romakloster, Sweden, [6]Moredun Research Institute, Midlothian EH26 0PZ, United Kingdom, [7]Royal GD, 7400 AA Deventer, The Netherlands

Bovine tuberculosis (BTB) and paratuberculosis (Johne's disease, JD) are production limiting diseases of confirmed/proposed zoonotic significance. In a recent survey of 48 countries, 38 had a control program for BTB while 22 had one for JD. The gap in control of JD was associated with the absence of OIE guidelines. Control of both diseases is problematical due to movements of infected livestock, imperfect diagnostic tests, lack of vaccines and lengthy control periods. Nevertheless, the principles of control for both diseases are similar and require on-farm biosecurity, individual animal diagnostic effort (property identification, animal identification, farm visits, sample collection, laboratory testing, coordinated record keeping/analysis) and action (culling). The heavy burden of BTB and JD identified in some countries suggests that promoting JD control alongside BTB control could boost awareness, enable efficient integrated surveillance and use a common diagnostic pipeline. Control of BTB has been conducted successfully together with eradication of bovine brucellosis (BTEC campaigns). Is there a case for joint control of BTB and JD, and what constraints exist to such an idea? A disincentive in many countries is that farmers would object to identifying a second category of mycobacterial reactors (JD) in their herds and resist this being actionable. However, maintaining compulsory actions for BTB reactors and voluntary approaches for JD may not be incompatible, would make efficient use of limited resources in both public and private sectors and may enable market and food processor driven initiatives to expand. Hurdles to be addressed by the research community include: reducing interference of JD vaccines and other mycobacterial antigen exposures with BTB diagnostics; reducing interference of BTB diagnostics (intradermal test) with MAP diagnostics; identifying efficacious vaccines for both diseases for use in high prevalence situations or where test and cull is impossible; and identifying the socioeconomic and geopolitical constraints to disease control.

298

Isolation of Novel Lytic Bacteriophages with the Potential to be Used for Detection of Viable Mycobacterium avium subsp. paratuberculosis

Miss Hannah Dane[1], Dr Linda Stewart[1], Professor Irene Grant[1]
[1]Institute for Global Food Security, School of Biological Sciences, Queen's University Belfast, Belfast, United Kingdom

There has been considerable interest in phage-based detection of viable Mycobacterium avium subsp. paratuberculosis (MAP) over recent years because test results are obtained much more quickly than by culture. The aim of this study was to isolate novel lytic mycobacteriophages capable of infecting and lysing MAP cells that may have the potential to replace the D29 mycobacteriophage currently employed in such tests. Five environmental samples (soil, lake water and bovine faeces) were tested for the presence of lytic mycobacteriophages, directly and after enrichment. Initial plaque assays were performed with a lawn of fast-growing Mycobacterium smegmatis, rather than MAP. Lytic phages were isolated from one enriched soil sample, and 10 plaques were randomly selected and individually amplified for further characterisation. Restriction enzyme digestion of phage DNAs with BamH1, Cla1, EcoR1, HaeIII and HindIII revealed that there may be up to four phage types amongst the 10 isolated. The host range was determined by performing plaque assays with phage cocktails and 13 different Mycobacterium spp. Plaques were observed on lawns of M. smegmatis, M. fortuitum and M. marinum. Ability to infect and lyse slow-growing MAP cells could not be determined by plaque assay, so was demonstrated by coating magnetic beads with selected phage isolates and carrying out PhMS-PCR. A PCR product was obtained after phagomagnetic separation of a $3x10^4$ cfu/ml MAP solution and incubation for 1 hour at 37°C. Potentially several different lytic mycobacteriophages have been isolated and ongoing research is comparing their performance relative to the D29 mycobacteriophage. This will assess what benefits in terms of detection sensitivity or shorter time to results, if any, they could provide if incorporated into phage-based assays to rapidly detect viable MAP.

301

The silent sleep of intramacrophage Mycobacterium avium subspecies paratuberculosis

Dr Joanne Stevens[1], Dr Kirsty Jensen[1]
[1]*University Of Edinburgh, ,*

Johne's disease is a huge economic and animal welfare issue in the livestock industry and new methods of detection and control are needed. The development of these requires a greater understanding of the interaction between the causative agent, Mycobacterium avium subspecies paratuberculosis (MAP) and bovine macrophages, where MAP resides. We have investigated the interaction between bovine monocyte-derived macrophages (bMDM) and three strains of MAP: the reference strain K10 and two field isolates (C49 and L1). Interrogation of a publicly available RNASeq dataset (GEO GSE104211) revealed that MAP strains induced a much more silent bMDM response than infection with Mycobacterium bovis strains. RT-qPCR analysis of several immune-related genes, e.g. IL1B, has confirmed this observation, with MAP infection inducing a more silent response than BCG, the attenuated vaccine strain of M. bovis. Whilst the overall magnitude of response was lower than for M. bovis infection, MAP strain specific variation in the bMDM response was observed, with K10 inducing the greatest response. In addition, the survival of MAP in bMDM was investigated by quantifying colony forming units (CFU) and genome copy number (GCN). CFU analysis revealed that the initial uptake and survival of the three MAP strains in bMDM differed significantly. However, by six days post infection this strain effect had disappeared and only approximately 15% of each of the MAP strains were recoverable. In contrast, quantification of MAP GCN found a much smaller decrease in MAP numbers suggesting that the MAP strains were still present. Further work using BacLight live/dead staining showed that a similar proportion of MAP were alive at 1 hour and 5 days post infection, implying that a large proportion of MAP are surviving in macrophages in a dormant, non-replicating state. This dormant state allows MAP to reside undetected within the very immune cell that should be destroying it.

303

Factors affecting the likelihood of a positive or negative Mycobacterium avium subspecies paratuberculosis (MAP) faecal PCR result in ELISA seropositive or inconclusive animals in Irish herds – preliminary results

Miss Niamh Field[1], Dr. John Mee[1], Dr. Conor McAloon[2]
[1]Teagasc, Fermoy, Ireland, [2]University College Dublin, Dublin 4, Ireland

The objective of this study is to determine the factors affecting the outcome of a faecal PCR test in animals that previously tested MAP ELISA-positive or inconclusive, and here we report on preliminary analysis relating to ELISA titre and PCR test results.

Materials and methods: ELISA (n=1,045,021) and PCR (n=13,608) results for all herds registered in the Irish Johne's Control Programme (IJCP) were obtained for 10 years (2012-2021). Data cleaning, processing and analysis was completed using R. The analysis was focused on animals that tested ELISA positive or inconclusive, as the IJCP requires ancillary faecal PCR testing to confirm infection in these animals. Animal-level (ELISA titre, parity, lactation length, TB test date), herd-level (apparent within-herd prevalence (aWHP), herd infection status) and test-level factors (ELISA sample type) were included in a logistic regression model with PCR test outcome (positive or negative) as the dependent variable.

Results: Only results for one factor, ELISA titre, are presented. Further modelling and analysis is continuing to determine the significance of other factors, with full results expected in 2022.There was a significant association between ELISA titre and the probability of a positive PCR result. However, even with relatively high ELISA titres (>200% S/P), the predicted probability of a PCR positive was still relatively modest (30-40%). Further modelling and analysis is continuing to determine the significance of other factors, with full results expected in 2022.

Conclusions: In these preliminary results, it is apparent that while ELISA titre is significantly associated with a PCR test result, the probability of a positive PCR result remains low even at relatively high titres. Further modelling is ongoing to investigate a range of factors that may affect the outcome of a PCR test for MAP.

305

Performance of the molecular diagnosis using qPCR IS900 for the detection of Mycobaceterium avium subspecies paratuberculosis in cattle feces in Chiriquí, Panama

Dr. Fermin Acosta[1], Lic. Daibellys De La Hoz[1], Dr. Carolina De La Guardia[1], MSc Ana Paula De Souza, MSc Dilcia Sambrano[1], Dr. Anabel García, Student Daniela Candanedo[6], Student Dayan Palacios[7], Student Sara Madrid[7], MSc Venancio Polanco[2], Dr. Andrea Prichard[2], Dr. Aristides Villarreal[4], Dr. Marcelino Jaén[4], Dr. Juan Carlos Pitti[3], Dr. Carlos Nader[3], **Researcher Amador Goodridge[1]**

[1]Instituto de Investigaciones Científicas y Servicios de Alta Tecnología (INDICASAT-AIP), City of Knowledge, PANAMA, [2]Universidad Tecnológica (OTEIMA), David, PANAMA, [3]Regional de Potrerillos, Ministerio de Desarrollo Agropecuario (MIDA), Potrerillos, PANAMA, [4]Centro de Innovación Agropecuaria de Chiriquí, Instituto de Innovación Agropecuaria de Panamá, David, PANAMA, [5]Laboratorio Veterinario LANAVET, Belo Horizonte, BRASIL, [6]Universidad Latina de Panama, Panama, PANAMA, [7]Universidad Autonoma de Chiriquí, David, PANAMA

Quantitative PCR is the most efficient method for detecting and quantifying the Mycobacterium avium subspecie paratuberculosis (MAP) burden in livestock. Our objective was to adapt and evaluate the performance of total feces DNA extraction and real-time PCR detection of IS900 from MAP. We prepared a positive internal control using an IS900 segment cloned from a reference strain MAP K-10 into a pUC19 plasmid. We then evaluated DNA extraction methods, including magnetic beads and hexadecyltrimethylammonium bromide (CTAB). Between January and November 2021, we obtained paired serum and fecal samples from 2,470 cattle and conducted commercial ELISA, qPCR and end-point PCR evaluation. Our results showed that the limit of detection (LOD) for the qPCR was nearly 0.008 ng of pUC19::IS900. When we used DNA from the MAP K-10 reference strain, the LOD was 0.001 ng, which is equivalent to 95 copies of the IS900 element. On the other hand, when using real-time PCR, the LOD threshold ranged from 7.78 to 33.99 Ct. When analyzing clinical samples, we found 2.7% with positive ELISA results among the animals under study. We noted that the organic method based on CTAB allowed DNA extraction with an average yield of 91.2 ng / ul (range 1.3 ng/ul – 2,009.2 ng/ul), while the magnetic bead-based method yielded 42.7 ng / ul (range= 0 ng/ul – 559.0 ng/ul). We found 34 (14.6%) had IS900 MAP DNA, while in the DNA extracted by CTAB we found 76 (32.6%) with IS900 MAP DNA. These results suggest that qPCR using the CTAB DNA extraction method has a better performance for the detection of IS900 MAP DNA in bovine fecal samples. We recommend using the CTAB DNA extraction method for bacterial dispersal surveillance along with the qPCR platform for future studies in Panama.

308

Single nucleotide polymorphisms genotyping of cattle from Uganda and Sudan and their connection to Mycobacterium subspecies paratuberculosis infection in Ugandan and Sudanese cattle

Dr Julius Okuni[1], Dr. Judah Ssekitoleko[1], Dr. Wisal Elmagzoub A[2], Dr. Sanaa Idris M.[3], Ms Monica Mpumwiire[1], Mr. Magid Kisekka[1], Prof, Lonzy Ojok[1], Prof Ahmad Amanzada[6], Dr. Elsagad Eltyayeb[4,7], Dr. Kamal Eltom[2], Dr. Ahmed Abd El Wahed[5]

[1]Department of Pharmacy, Veterinary Clinical and Companion Animal Medicine, College of Veterinary Medicine, Animal Resources and Biosecurity, Makerere University, Kampala, Uganda, [2]Unit of Animal Health and Safety of Animal Products, Institute for Studies and Promotion of Animal Exports, University of Khartoum, Shambat 13314, Khartoum North, Khartoum, Sudan, [3]Department of Pathology, Faculty of Veterinary Medicine, University of Khartoum, Shambat 13314, Khartoum North, Khartoum, Sudan, [4]Ibn Sina Specialised Hospital, Mohammed Najeeb st., Khartoum, Sudan, [5]Institute of Animal Hygiene and Veterinary Public Health, Faculty of Veterinary Medicine, University of Leipzig, An den, Tierkliniken 1, D-04103, Leipzig, Germany, [6]Department of Gastroenterology and Gastrointestinal Oncology, University Medical Centre Goettingen, Georg-August University Goettingen (UMG), Goettingen, Germany, [7]Al Neelain University, Khartoum, Sudan

Mycobacterium avium subspecies paratuberculosis (MAP) causes Johne's disease, a chronic debilitating disease in ruminants associated with an intractable diarrhea, production losses and death. Single nucleotide polymorphisms (SNPs) in a number of genes have been linked to genetic susceptibility or resistance is believed to play a significant role in the development of the disease following an infection with MAP. A small study we did previously had suggested that some of the SNPs were monomorphic while others showed significant association with sero-positivity in one breed of cattle in Uganda. The aim of our study was to compare the genotype and allele frequency of MAP positive and MAP negative cattle under the same management in a larger population of cattle. Currently, we have Partial data and we hope to complete the analysis within a month to present the detailed findings at the conference

309

Lessons from the Irish Johne's Control Programme

Lawrence Gavey[1], David Graham[1], Simon More[2], Lorna Citer[1]
[1]Animal Health Ireland, Carrick on Shannon, Ireland, [2]Centre for Veterinary Epidemiology and Risk Analysis, UCD School of Veterinary Medicine, University College Dublin, Ireland

The IJCP commenced in October 2017 following a four-year pilot programme to develop objectives, systems and processes. Almost 2,000 registered herds voluntarily participate.
The programme undergoes continual improvement to address emerging issues and enhance its value to participating herdowners and milk processors.
The development and implementation of the IJCP have been shaped and challenged by:

Context:
Building of trust and goodwill between stakeholders through open and constructive dialogue
Agreement on cost-sharing
Changing contexts of trade and politics
Competing but compatible objectives of supporting herds that are infected (estimated 30% herd prevalence) and those not infected
Building on lessons learned by other countries (acknowledging their generosity in sharing their experience)
Strengths and risks of a range of communication channels

Technical elements:
Balancing the technical complexities of Johne's disease and its control against stakeholders' needs for simple and cost-effective solutions
Balance of technical integrity and on-farm practicality

Implementation:
Multiple priorities and limited resources, particularly time, for farmers and veterinary practitioners
Reducing the complexity of recording and administering a JD programme to sets of computer algorithms
Compliance with annual testing and management plan requirements
Concerns about "test inaccuracy", complexity, protracted results, cost-effectiveness, and uncertain outcomes.
Constructive engagement with programme stakeholders has been key to resolving these issues.

312

Genetic typing of Mycobacterium avium subsp. paratuberculosis strains isolated from 14 cows shedding high levels of MAP in the Provinces of Quebec and Ontario Herds

PhD student Alexander Byrne[1], Kapil Tahlan[1], Research assistant Séverine Ollier[2], Research Assistant Thierry Cochard[3], Scientific Researcher Franck Biet[3], **Dr Nathalie Bissonnette[2]**

[1]Department of Biology, Memorial University of Newfoundland, St. John's, Canada, [2]2Sherbrooke Research and Development Center, Agriculture and Agri-Food , Sherbrooke, Canada, [3]INRAE, ISP, Université de Tours, Nouzilly, France

Mycobacterium avium subsp. paratuberculosis (MAP) is the causal bacteria for Johne's Disease, a chronic granulomatous enteritis affecting both wild and domestic ruminants. During the subclinical stage of infection, infected animals may excrete MAP in feces, allowing disease to spread undetected throughout the herd. Epidemiological studies have used PCR-based techniques to examine loci found throughout the genome known as MIRU-VNTR. Additional studies have used DNA fragment analysis performed on specific short sequence repeats (SSRs), which also discriminate strains. An important consideration is the occurrence of mixed strain infections (MSIs), which may exhibit different phenotypes or antigens, complicating the immune response. Instances of MSIs are typically considered as rare events, with few studies successfully identifying MSIs within MAP host animals. This study examines strains of MAP isolated from 14 cows and evaluates molecular methods for rapid MSI identification, notably using MIRU-VNTR and SSR-based fragment analysis. Cows were selected from three farms and were categorized based on blood ELISA and fecal ISMAP02 qPCR results. Strains of MAP (10 per cow) were selected on solid cultures. For each strain, an axenic liquid culture was prepared, totalling 139 isolates. DNA extracted from each strain was examined using 8 MIRU-VNTR, 2 SSR loci, and whole genome-sequencing. The MIRU-VNTR analysis results were identical across all isolates. However, the results from the SSR fragment analysis showed up to 9 unique patterns within each animal suggesting that MSIs may be present. As the rate of change of SSR could be independent of the rate of chromosomal evolution, further work will include more JD cases, which provides the greatest resolution for differentiation between strains. Results of this analysis show that SSR can discriminate strains with greater resolving ability than MIRU-VNTR. Our study suggests that MSI events may be overlooked within herds and should be considered in a phylogenetic context for affirming epidemiological connections.

313

Bayesian estimation of sensitivity and specificity of fecal culture, fecal PCR and serum ELISA for diagnosis of Mycobacterium avium subsp. paratuberculosis infections in Egyptian sheep

Dr Juan Carlos Arango Sabogal[1], Dr Ibrahim Elsohaby[2,3], Dr Abdelfattah Selim[4]

[1]Département de pathologie et microbiologie, Faculté de médecine vétérinaire, Université de Montréal, Saint-hyacinthe, Canada, [2]Department of Animal Medicine, Faculty of Veterinary Medicine, Zagazig University, Zagazig City, Egypt, [3]Department of Health Management, Atlantic Veterinary College, University of Prince Edward Island, Charlottetown, Canada, [4]Department of Animal Medicine (Infectious Diseases), Faculty of Veterinary Medicine, Benha University, Toukh, Egypt

Our objective was to evaluate the diagnostic accuracy of the individual fecal culture (IFC), fecal PCR (FPCR), and serum ELISA for the detection of Mycobacterium avium subsp. paratuberculosis (MAP) infections in sheep from four governorates in Egypt using a latent class model (LCM) fitted within a Bayesian framework (BLCM).

Fecal and blood samples were collected from 370 sheep in four Egyptian governorates. Fecal samples were analyzed by IFC and RT-PCR based on ISMav2 insertion sequence, while ELISA was performed on serum samples. The median sensitivity (Se) and specificity (Sp) and 95% Bayesian credibility intervals (95% BCI) of the three diagnostic tests were estimated using a hierarchical BLCM.

The median Se estimates for IFC, FPCR, and serum ELISA were 32.2% (23.3-42.2), 52.0% (33.2-84.2), and 64.2% (43.0-83.4), respectively. The median Sp estimates for IFC, FPCR, and serum ELISA were 97.7% (96.1-98.8), 97.6% (95.5-99.4), and 98.4% (96.8-99.3), respectively. The median within-governorate MAP prevalence was 11.9% (BCI: 5.0-27.8).

At a threshold of >45%, ELISA showed the highest Se and comparable Sp to IFC and FPCR. The test ELISA evaluated in this study is an interesting alternative for detecting MAP in sheep due to its higher Se, lower cost, and shorter turnaround laboratory time compared to IFC and FPCR.

314

Evaluation of associations between MAP-C or MAP-S genotypes and the severity of histo-/pathological abnormalities of infected cattle, red deer and sheep from Germany and New Zealand

Dr. Marian Price-carter[1], Dr. Milan Gautam[2], Dr Patrick Biggs[3,4], Dr. Michael Weber[5], Dr. Heike Koehler[5], Dr. Cord Heuer[2], Dr. Petra Mobius[5]

[1]Agresearch Ltd, Palmerston North, New Zealand, [2]EpiCentre, School of Veterinary Science, Massey University, Palmerston North, New Zealand, [3]School of Fundamental Sciences, Massey University, Palmerston North, New Zealand, [4]Molecular Epidemiology and Public Health Laboratory, School of Veterinary Science, Massey University, Palmerston North, New Zealand, [5]Institute of Molecular Pathogenesis, Friedrich-Loeffler-Institut, Jena, Germany

If virulence differences exist between Mycobacterium avium subspecies paratuberculosis (MAP) subtypes, they may be attributed to genetic characteristics that could be exploited in tools used to control paratuberculosis (Ptb) in farmed ruminants. Currently, we have no confirmatory evidence of virulence differences between MAP genotypes. Circumstantial evidence from population-based studies in New Zealand suggest that certain genetic variants of MAP might be associated with different pathogen virulence and a preliminary investigation in Germany found that IS900-RFLP-(BstEII)-Type-C1 was isolated more often in cattle with pathological macroscopic lesions characteristic of Ptb compared to those without lesions. The current study aimed to uncover associations between phenotypical and genetic characteristics of individual MAP strains originating from different host species and environments. We investigated genomic differences between MAP isolates from German cattle with and without Ptb typical macroscopic lesions within the gut, and New Zealand deer and sheep with phenotypic tissue markers (histopathology), clinical signs, and varying shedding levels. In addition, isolates from before and after passage through cattle and sheep hosts were studied. All of these data were subjected to conventional and WGS bioinformatics methods including WGS-SNP-profiling, combined MIRU-VNTR, IS900-RFLP and SSR genotyping. We did not observe any association between specific MAP genotypes and phenotypes in any of the examined hosts (cattle and deer and sheep) or MAP sub-types (MAP-C, sheep MAP-S). However, results of WGS-SNP analysis showed a strong association with that of combined genotyping. Furthermore, no additional SNPs were detected in MAP strains after passages thru different animal hosts. In conclusion, results likely reflect the slow rate at which MAP incorporates changes into its genome and suggest that differences in clinical signs in MAP infected animals are more likely to be caused by pathogen independent characteristics such as variable infective dose, frequency of, and age at exposure, and variation in host immune response.

AUTHOR INDEX

Aalberts, M. 18, 39, 93, 96
Acosta, F. 129
Agarwal, P. 87
Agrawal, G. 116
Alfituri, O. 2
Allen, A. 3
Alonso, M. 32, 49, 101
Alonso, N. 13
Alonso-Hearn, M. 35, 53, 56, 58
Alston, L. 54
Amado, J. 32, 35
Amanzada, A. 22, 78, 130
Arrigoni, N. 60, 72, 81
Aseri, G. 34, 63, 105, 106, 107, 114
Attili, A.-R. 120, 124
Aubock, L. 17
Aulik, N. 89, 90
Bach, H. 62
Badia-Bringue, G. 32, 49, 53, 56, 58
Badiola, A. 104
Bakony, M. 73
Balseiro, A. 35, 122
Bannantine, J. 6, 12
Barbosa, W. 9
Barkema, H. 15, 31, 67
Barsi, F. 72, 81
Baruta, G. 54
Bauman, C. 21, 29
Baumgartner, W. 98
Bay, S. 62
Beaulieu, E. 67
Beaunee, G. 69
Behr, M. 8, 14, 80
Beiting, D. 10
Bertazzo, V. 38
Bhat, A. 11
Biagetti, M. 99
Bianchi, A. 72
Biet, F. 6, 12, 62, 85, 132
Biggs, P. 134
Bilbao, J. 58
Bissonnette, N. 57, 132
Blake, R. 2
Blanco-Vazquez, C. 32, 35, 56, 58
Bonacucina, E. 120, 124
Bond, K. 66
Bottazzari, M. 38
Branger, M. 6, 12
Brasil, A. 92
Bremaud, I. 91
Brown, S. 121
Browne, J. 3
Buczinski, S. 61
Burrell, A. 19
Byrne, A. 132
Caldeira, J. 9
Caldow, G. 33
Cammi, G. 81
Campe, A. 23, 83
Candanedo, D. 129
Canive, M. 32, 35, 49, 53, 56, 58
Cannon, R. 123
Cano, P. 51
Cardoso, F. 102
Casais, R. 32, 35, 56, 58, 122
Cassidy, S. 3
Chang, Y.-F. 9
Chaubey, K. 11, 44, 59, 74, 87, 110, 112
Chauhan, D. 74, 112
Ciordia, S. 122
Citer, L. 24, 26, 108, 131
Ciullo, M. 99
Clegg, T. 123
Cochard, T. 6, 12, 85, 132
Conde, C. 6, 12
Conley, K. 90
Cooper, J. 89, 90
Corbiere, F. 68
Corlatti, L. 72
Corneli, S. 99
Corrales, F. 122
Cosciani-Cunico, E. 81
Cossu, D. 100
Costarelli, S. 99
Cote, G. 61
Cournet, J. 68
Crispell, J. 3
Crotta, M. 66
Cuerda, X. 101
Cuteri, V. 120, 124
Dahmen, H. 27
Dalzini, E. 81
Daminelli, P. 81
Dane, H. 36, 126
Davergne, A. 6
Davies, C. 103

De Buck, J. 31, 38, 52, 54
De La Guardia, C. 129
De La Hoz, D. 129
De Silva, K. 55
De Silva, Kumudika 47
De Souza, A. 129
Delafosse, A. 6
Di Paolo, A. 99
Diaz-Miranda, E. 9
Dijkstra, T. 18, 39, 93, 96
Djitro, N. 7
Donat, K. 23, 70, 83, 84, 86, 111
Dore, E. 61
Dubaux, E. 68
Duffy, S. 8, 14
Dunser, M. 17, 76
Earle, E. 42
Egan, J. 123
Einax, E. 70, 86
Eisenberg, S. 23, 83, 125
Elguezabal, N. 104
Elmagzoub, W. 22, 130
Elsohaby, I. 133
Eltayeb, E. 22
Eltom, K. 22, 78, 130
Eltyayeb, E. 78, 130
Engemann, C. 88
Evans, C. 45, 46
Ezanno, P. 69
Facciuolo, A. 48, 51
Farrell, D. 3
Farrell, E. 108
Fecteau, G. 57, 61
Fecteau, M.-E. 10
Fernandez, A. 49
Fernandez, N. 58
Field, N. 64, 128
Filippi, A. 60, 72, 81
Filippini, G. 99
Fiorini, P. 38
Foddai, A. 42, 121
Fornyos, K. 98
Forti, A. 72
Foucras, G. 68
Foures, L. 68
Fourichon, C. 6
Fuchs, T. 115
Gabbianelli, F. 99
Galleti, G. 60
Galosi, L. 124
Gandhi, H. 63
Ganneau, C. 62
Garbarino, C. 60, 72, 81
Garcia, A. 129
Garcia, V. 103
Garrido, J. 35, 49, 58
Gaunitz, C. 88
Gautam, A. 44, 59, 87
Gautam, M. 134
Gavey, L. 19, 24, 26, 108, 131
Geraghty, T. 33
Gerdts, V. 48, 51
Gieseke, A. 5
Goethe, E. 5
Goethe, R. 5
Gonzalez, A. 62
Gonzalez-Recio, O. 49
Good, M. 19
Goodridge, A. 129
Gordon, N. 28, 40
Gordon, S. 3
Graham, D. 19, 24, 26, 108, 131
Grant, I. 25, 36, 41, 42, 121, 126
Gravisaco, M. 101
Green, M. 66
Greenstein, R. 121
Griebel, P. 48, 51, 57
Griffin, F. 65, 75, 125
Griffin, M. 123
Guatteo, R. 6
Guellouz, D. 68
Guimaraes, J. 9
Guitian, J. 66
Gunaseelan, K. 74
Gupta, S. 11, 34, 59, 74, 87, 105, 106, 107, 112
Hall, D. 67
Hanafy, M. 50
Hancock, R. 48, 51
Hand, D. 123
Handley-Hartill, W. 46
Harihahan, S. 44
Hariharan, S. 110
Hartill, W. 45
Hechard, Y. 85
Hermon-Taylor, A. 116
Hermon-Taylor, J. 116
Heuer, C. 134
Hirota, S. 54
Hodgeman, R. 7
Hope, J. 2
Hossein, S. 100
Hudspith, B. 116
Huerta, B. 102
Hunyh, A. 45

Hussain, T. 11
Ibeagha-Awemu, E. 57
Idris, S. 22, 130
Iglesias, N. 32, 35, 122
Iglesias, T. 35
Imada, J. 21, 29
Jaen, M. 129
Jain, M. 34, 63, 105, 106, 107, 114
Janeczko, R. 28, 40
Jasemi, S. 100
Jensen, K. 2, 127
Jessu, A. 85
Joly, A. 6, 69
Journaux, L. 6
Junior, A. 9
Jurado-Martos, F. 102
Jurkovich, V. 73, 98
Juste, R. 32, 35, 49, 58
Kapur, V. 30
Kasbohm, E. 86
Kelton, D. 21, 29, 57
Kenny, K. 3
Khol, J. 17, 76
Kisekka, M. 130
Klassen, A. 70
Koehler, H. 109, 115, 134
Kohler, H. 84, 111
Kossler, J. 17
Krieger, M. 23
Kubala, A. 46
Kumar, A. 34, 63, 105, 106, 106, 107, 114
Laarmann, K. 5
Labitzke, M. 88
Ladero, I. 104
Lavin, J. 56, 58
Lee, A. 48, 51
Leon, A. 122
Liebler-Tenorio, E. 115
Liening-Ewert, T. 17
Linardi, M. 120, 124
Links, I. 27
Lorenz, I. 23, 94
Losio, N. 81
Luhrs, J. 5
Lupien, A. 14
Luque, I. 102
Luzzago, C. 72
Mader, C. 17
Madrid, S. 129
Malik, H. 11
Mann, R. 7
Markey, B. 123
Martins, L. 31
Mason, S. 67
Massaro, A. 38
Maurya, S. 114
Mazzone, P. 99
McAloon, C. 19, 64, 128
McElroy, S. 108
McInerney, B. 42
McIntosh, F. 14
Mee, J. 64, 128
Meek, K. 25, 41
Meloni, G. 100
Michel, A. 125
Milhano, F. 122
Mir, A. 11
Mobius, P. 109, 134
Moebius, P. 115
Molina, E. 104
Moog, U. 111
Moore, H. 28
More, S. 19, 26, 131
Moreira, M. 9, 92
Moroni, M. 95
Mourya, S. 63
Moussi, L. 88
Moyano, R. 13, 101
Moyen, J.-L. 85
Mpumwiire, M. 130
Mujika, M. 104
Munoz, M. 122
Mutharia, L. 51
Mutlow, A. 90
Nader, C. 129
Nagel, A. 13, 101
Napper, S. 48, 51
Narang, P. 114
Narang, R. 114
Nava, M. 72
Ngwa, V. 120, 124
Ngwa-Mbot, D. 6
Nicolas, P. 69
Nighallchoir, E. 123
Noli, M. 100
Norton, S. 75
Nunney, E. 66
O'Connell, L. 43
O'Connell, N. 25, 41
O'Connor, A. 3
O'Mahony, J. 3, 43
O'Sullivan, P. 24
Oehlers, S. 55
Ojok, L. 22, 78, 130

Okuni, J. 22, 78, 130
Oliveira, L. 92
Olivieri, M. 13, 101
Ollier, S. 132
Omega, J. 22
Orpin, P. 16, 20, 71, 97, 117, 118, 119
Orsel, K. 31
Ortner, P. 17
Oyanguren, M. 104
Pacifici, L. 120, 124
Palacios, D. 129
Pallante, I. 38
Pena, J. 9, 92
Pendrotti, L. 72
Perehinec, T. 45
Perets, V. 3
Petrucci, L. 99
Pezzotti, G. 99
Pickrodt, C. 84, 109, 111, 115
Pickup, R. 116
Pineau, L. 91
Pinedo, F. 101
Piro, R. 38
Pitti, J. 129
Plain, K. 55, 103
Polanco, V. 129
Polavarapu, R. 34, 105, 106, 107
Porteen, K. 74
Potter, A. 48, 51
Pozzato, N. 38
Pra, M. 38
Prendergast, D. 123
Price-Carter, M. 1, 12, 75, 134
Prichard, A. 129
Pryor, S. 79, 82
Purdle, A. 55
Queipo, M. 35
Ramovic, E. 123
Rana, S. 114
Rankin, J. 42
Rasmussen, P. 67
Rawlyk, N. 48
Rayment, N. 116
Reddy, J. 34
Rees, C. 45, 46
Rhodes, G. 116
Ricchi, 72
Ricchi, M. 60, 72, 81
Richardson, G. 33
Roche, S. 21, 29
Rochfort, S. 7
Rodoni, B. 7
Romano, M. 13, 101
Rossignol, M.-N. 6
Roy, J.-P. 61
Russo, S. 60, 72, 81
Sabogal, J. 61, 133
Salgado, 37
Salgado, M. 77, 95, 113
Samani, R. 52
Samba-Louaka, A. 85
Sambrano, D. 129
Sanderson, J. 116
Sasse, O. 88
Savin, K. 7
Sbarra, F. 99
Schibler, L. 6
Schrott, J. 76
Schukken, Y. 18, 39, 93, 96
Schwan, E. 125
Schwarz, D. 92
Scoccia, E. 99
Sebastianelli, M. 99
Sebastiani, C. 99
Sechi, L. 100
Selim, A. 133
Sharma, D. 34, 63, 105, 106, 107, 114
Sharma, S. 44, 59
Sibley, D. 16, 20, 117, 118, 119
Sibley, R. 71, 97
Silva, A. 92
Singh, A. 34, 74, 105, 107, 112
Singh, J. 63, 114
Singh, M. 34, 87, 105, 106, 112
Singh, P. 74, 87, 110, 112
Singh, S. 11, 34, 44, 59, 74, 87, 110, 112
Singh, V. 106
Skuce, R. 3
Slavin, Y. 62
Smith, J. 89
Sodoma, E. 76
Sohal, J. 34, 74, 105, 106, 107
Sollner-Donat, S. 70
Sosa, P. 101
Ssekitoleko, J. 22, 78, 130
Stabel, J. 121
Stefani, A. 38
Steuer, P. 37, 77, 95, 113
Stevens, J. 2, 127
Stevenson, K. 12, 125
Stewart, L. 36, 126
Strain, S. 25, 41, 42
Su, L. 121
Sweeney, R. 10

Swift, B. 45, 46
Tabouret, M. 91
Tahlan, K. 57, 132
Talaat, A. 50, 90
Tandon, N. 88
Tanneberger, F. 22
Tarradas, C. 102
Tasca, C. 68
Tata, A. 38
Taylor, N. 117
Tejea, C. 95
Tejeda, C. 37, 77, 113
Tejerina, J. 28, 40
Thibault-Poisson, V. 91
Thirion, M. 91
Thomas, A. 103
Thresher, R. 79, 82
Tichy, A. 17, 76
Toribio, J.-A. 103
Torricelli, M. 99
Townsend, H. 48
Traveria, G. 101
Ullman, F. 88
Valentini, A. 99
Van Winden, S. 66
Vass-Bognar, B. 73, 98
Vazquez, C. 122
Vazquez, P. 35, 49, 58
Verdugo, C. 95
Vergu, E. 69
Villarreal, A. 129
Villegas, M. 37, 77, 113
Wahed, A. 22, 78, 130
Wang, Y. 4
Watson, G. 42
Webber, M. 125
Weber, M. 18, 39, 93, 96, 134
Wellemans, V. 61
Wells, S. 4
White, R. 79, 82
Whittaker, D. 28, 40
Whittington, R. 12, 103, 125
Wichert, A. 86
Wright, K. 55
Yearsley, D. 123
Zhang, H. 54
Zilch, T. 9